Developments in the Call Centre Industry

The call centre industry is one of the fastest growing areas of employment in the world. Advances in information and communication technology (ICT) have led to much of the industry being automated and relocated from advanced OECD economies to countries with lower labour costs.

Developments in the Call Centre Industry draws on contributions from a diverse range of countries including the USA, the UK, India, Australia, South Korea, Germany, Greece and Sweden. Authors examine the emergence of the call centre sector as an international industry. Themes covered in this original volume include call centres and:

- the recruitment, training, motivation and retention of the workforce
- the influence of pressures on operatives and managers to reach operating targets
- gender and career aspects of the industry
- trade unions and the organisation of call centre workers.

John Burgess is an Associate Professor and Director at the Employment Studies Centre, University of Newcastle, Australia. **Julia Connell** is an Associate Professor at the Newcastle Graduate School of Business, University of Newcastle, Australia.

Routledge studies in business organizations and networks

Developments in the Call Centre Industry

Analysis, changes and challenges

Edited by John Burgess and Julia Connell

 Routledge
Taylor & Francis Group

LONDON AND NEW YORK

First published 2006
by Routledge
2 Park Square, Milton Park, Abingdon, Oxon OX14 4RN

Simultaneously published in the USA and Canada
by Routledge
270 Madison Ave, New York, NY 10016

Reprinted 2007

Routledge is an imprint of the Taylor & Francis Group, an informa business

© 2006 Selection and editorial matter, John Burgess and Julia
Connell; individual chapters, the contributors

Typeset in Times by Wearset Ltd, Boldon, Tyne and Wear
Printed and bound in Great Britain by TJI Digital, Padstow, Cornwall

British Library Cataloguing in Publication Data
A catalogue record for this book is available from the British Library

Library of Congress Cataloging in Publication Data
A catalog record for this book has been requested

ISBN10: 0-415-35702-0 (hbk)
ISBN10: 0-203-00300-4 (ebk)

ISBN13: 978-0-415-35702-9 (hbk)
ISBN13: 978-0-203-00300-8 (ebk)

Contents

Figures

Tables

Notes on contributors

Peter Bain is Senior Lecturer in the Department of HRM, University of Strathclyde, Glasgow. Areas in which he has published include occupational health and safety, workplace technological change and contemporary developments in trade unionism. A lead applicant in a Scottish universities' project under the ESRC's 'Future of Work' programme, he has also studied work and employment relations in call centres in the UK, USA, the Netherlands and India.

John Burgess is an Associate Professor in the Newcastle Business School, University of Newcastle, Australia. He is also Director of the Employment Studies Centre, Faculty of Business and Law. Research interests include contingent employment arrangements, gender and work, and labour market policy. Recent publications include *International Perspectives on Temporary Agency Work* (with Julia Connell), Routledge, London, 2004.

Julia Connell is an Associate Professor at the Newcastle Graduate School of Business, Australia. Other than call centres, research areas include contingent employment arrangements, organisational and individual effectiveness – organisational culture, organisational change, management style, learning and skill development within various workplace settings and situations.

Gail Drummond was lead organiser with the Community and Public Sector Union for six years where her focus was predominantly phone companies and outsourcing. She then worked as an organiser and educator for the Australian Council of Trade Unions, before beginning her current position at the Health and Community Services Union.

Susan Durbin is a Lecturer in the School of Human Resource Management, University of the West of England, Bristol, United Kingdom. Susan's research interests include women in management, the gendering of new employment forms (call centres), the knowledge economy and the gender–technology relationship.

Maeve Houlihan is a Lecturer in Organisational Behaviour at UCD Business Schools, University College Dublin, Ireland. She holds a doctorate from the Department of Organisation, Work and Technology at Lancaster University. Her research focuses on frontline service work, contemporary working lives and management practices and their links with society. She is currently researching the Irish call centre sector as part of the Global Call Centre Project led by Cornell and Sheffield Universities.

Hye-Young Kang is a Research Fellow in the POSCO Research Institute (POSRI), Korea. Her recent research focuses on employment relations of the telecommunications service sector, job appraisal of shopfloor management, innovative compensation plans and labour relations reforms in the steel industry.

Aikaterini Koskina is a Lecturer in HRM at the Business School University of Sunderland and a doctoral student in the Department of HRM and Industrial Relations at Keele University, where she has also been a Graduate Teaching Assistant. Her current research focuses on the HRM policies and practices of European multinational corporations; and the nature and management of customer service work, with particular reference to the impact of managerial control on the organisation of work.

Byoung-Hoon Lee is an Associate Professor in the Department of Sociology, Chung-Ang University, Korea. His recent research interests include non-standard labour issues, labour market segmentation, the impact of information-communication technology on work and employment relations, reforms of labour union movement, social dialogue and workplace innovation.

Antoni Lindgren is a Senior Lecturer and Associate Professor in sociology at Luleå University of Technology, Sweden. He currently lectures in sociological theory, especially the classics, and the philosophy of science. His research is in the new service work and vocational education and training. Recent publications include: Anja Heikinen and Antony Lindgren (eds): *Social Competence in Vocational and Continuing Education*, Peter Lang, Bern 2004 and Antoni Lindgren: *Inidividen och samhället* (The Individual and Society), Bokförlaget i Baktsjaur, Baktsjaur 2005.

Al Rainnie is a Professor in the Department of Management at Monash University, Victoria; he is also Director of the Monash Institute for Regional Studies. Al has researched and written widely on the subjects of industrial relations and small firms, globalisation, regional development and the restructuring of work and organisations.

Bob Russell is an Associate Professor in the Department of Management at Griffith University. Prior to that, he worked for many years in the Sociology Department at the University of Saskatchewan. His previous publications include *More with Less*: University of Toronto Press, 1999, which examines work restructuring in industry. Bob is currently working on an Australian Research Council grant 'Voice, Representation and Recognition in the Information Economy' which explores issues of employee identity and union organising in contact centres.

Per Sederblad is Researcher and Senior Lecturer at Malmö University. His publications include: 'The Swedish Model of Work Organization in Transition' (with Paul Thompson in *Global Japanization? The Transnational Transformation of the Labour Process* edited by Tony Elger and Chris Smith, Routledge, 1994) and 'Teamworking and Emotional Labour' (with Antoni Lindgren in *Learning to be Employable. New Agendas on Work, Responsibility and Learning in a Globalizing World* edited by Christina Garsten and Kerstin Jacobsson, Palgrave, 2004). His current research focuses on team working and professions in the service sector.

Snigdha Srivastava is a PhD candidate in the Urban Planning and Policy programme at the University of Illinois at Chicago and a Researcher at the Center for Urban Economic Development. Her research interests involve transnational organisation of service jobs, knowledge industries and workforce development in the information technology industry.

Phil Taylor is Professor of Work and Employment Studies in the Department of Management and Organization at the University of Stirling. He has researched and published extensively on call centres, trade unions, occupational health and work organisation. He is also a lead member of an English Social Science Research Council 'Future of Work' project.

Nik Theodore is Assistant Professor of Urban Planning and Policy and Director of the Center for Urban Economic Development at the University of Illinois at Chicago. His research focuses on labour market restructuring, urban inequality, contingent work and employment policy.

Claudia Weinkopf is a Research Director at the Institute for Work and Technology in Gelsenkirchen, Germany, and Coordinator of the research unit 'Flexibility and Security'. Her research interests include labour market policy and employment issues – particularly with regard to services, atypical employment and low wages.

Foreword

Call centres have attracted considerable interest in the media and more recently amongst academic researchers. This is principally due to the nature and character of the jobs that have been created and to their spectacular growth and importance as a source of employment in many industrialised countries such as Australia. Call centres have been variously described as 'electronic sweatshops' and 'human battery farms', characterised as providing dead-end jobs that are poorly paid, closely monitored and highly routinised. Although this is far from a uniform picture it is clear that the purpose of call centres is to deliver customer service at the lowest possible cost. They act to centralise service provision and to consolidate service providers. Call centres rationalise the work process through the extensive use of advance information technology and they seek to standardise service encounters with functionally equivalent and interchangeable service providers.

This has important implications for worker well-being. The pressure to maximise call volume and minimise costs can lead to jobs that limit employee discretion and fail to make full use of workers' skills. But this need not be the case. Call centres can provide services through a different type of employment model where work can be designed to facilitate autonomy. Investment can be made in training and development, teamwork can be emphasised and supportive supervision can be nurtured. Furthermore, tightly scripted conversational rules can be relaxed and computer monitoring can be used for developmental rather than punitive purposes. Moreover, workers can be supplied with a variety of tasks that relieve them from the constant pressures of customer interactions.

Some call centres have moved towards a more empowered employment model. Sometimes this has been in response to pressure from customers who are disenchanted with the quality of the service they are receiving. An emphasis on service throughput to the detriment of service quality can erode customer loyalty and damage the organisation's reputation for competence in service delivery. Pressure can also come from employees who are emotionally drained by the repetitious nature of the job and dissatisfied with the arduous working environment they endure. High levels of stress and burnout can spill over into absenteeism and turnover and force management to rethink the logic of their work regime. In other cases, unions have helped to voice collectively the discontent of workers and secure important improvements in job design and in systems of management control. There have also been instances of workers turning the tables on their organisation by using management's espoused values of care and respect for the customer to defend and extend their own rights to fair and respectful treatment and acceptable working conditions.

However, this is not widespread. Most call centres bear the hallmarks of an engineering model and are run along the lines of a production line. Jobs are narrowly constructed, interactions with customers are tightly scripted and electronic surveillance is widespread. Employees often have very little free space during their working day and are presented with few opportunities for an amnesty from the constant emotional demands of the job. This lack of porosity in the working day can also rob unions of the time necessary to secure the involvement of employees in day-to-day union activities. Moreover, the adoption of what has been termed 'sacrificial HR strategies' whereby call centres are willing to sacrifice the well-being of staff and accept emotional burnout and high turnover as the price of maintaining high levels of service at low cost can lead to a continual haemorrhaging of union members and a constant pressure to recruit new members. The outsourcing and offshoring of call centre work to developing countries in eastern Europe and in Asia also present formidable challenges for trade unions. The growth of large multinational service providers and the increasing mobility of capital, facilitated by falling IT and telecommunication costs, have the potential to undermine the wages and working conditions of call centre workers in advanced economies such as Australia.

This collection of papers edited by John Burgess and Julia Connell is to be warmly welcomed. It deals with a wide range of important matters relating to the location (and offshoring) of call centres, employment arrangements in different countries, union organisation, and the nature and skills of the work and career opportunities for female workers. A consideration of these issues is critical to an understanding of call centre work: an area of employment activity that has become emblematic of the rapidly growing global service market. As a result I am sure that this book will be widely read and extensively cited.

Professor Stephen Deery[1]
Department of Management
King's College,
University of London,
United Kingdom

Note

1 Co-editor of: Deery, S. and Kinnie, N. (eds) *Call Centres and Human Resource Management: A Cross-national Perspective*, Palgrave Macmillan, Houndmills: 2004.

Acknowledgements

The editors would like to thank the University of Newcastle Research Grants Committee for supporting their research on call centres. In organising the book we relied heavily on the editorial assistance of Kate Flint, without whom we would never have reached publication stage. Thanks also go to Robert Langham at Routledge for his ongoing support and encouragement. Finally, the support of Tessa, Peter, Alex, Laura and Jonathan was crucial in finishing this project.

Copyright permissions

Abbreviations

ABS	Australian Bureau of Statistics
ACD	Automatic call distribution
ACTU	Australian Council of Trade Unions
BMS	Building management systems
BPO	Business process outsourcing
CBD	Central business district
CEO	Chief executive officer
CHT	Call handling time
CIO	Chief information officers
CPSU	Community and Public Sector Union
CRM	Customer relations manager
CSR	Customer service representatives
ECRM	Electronic customer relations manager
EIAS	Greek Institute of Training for Insurance Services
EIRR	European Industrial Relations Report
FDI	Foreign direct investment
GCCB	Global call centre benchmarking
HPWS	High performance work systems
HR	Human resources
HRM	Human resource management
HSBC	Hong Kong and Shanghai Banking Corporation
ICT	Information and communication technology
IT	Information technology
ITES	Information technology enabled services
IVR	Interactive voice response
KPI	Key performance indicators
MNC	Multinational corporation
OECD	Organization for Economic Co-operation and Development
SEC	State Electricity Commission
SLA	Service level agreement
TNC	Trans-national corporation
TQM	Total quality management
UNI	Union Network International
VoIP	Voice on Internet protocol
VRU	Voice recognition unit

tions can reduce their core employee numbers and costs, while still bene-fiting from continuous, and in some cases, extended service provision. Indeed, one could say that the impact of the 1980s and 1990s public sector restructuring in many Anglo-Saxon countries (the UK, Australia and New Zealand) has demonstrated how organisations can be restructured and workforces reduced (see the case studies for Australia in Fairbrother *et al.* 2002). The possibilities offered by ICT, which are driven by the impera-tives of competition and shareholder value, are enabling organisations to undertake production in different ways, with different forms of organisa-tion and in different locations.

The third factor related to the growth in call centres, which is related to the other two, is that it captures how ICT and the externalisation of pro-duction have facilitated the restructuring of service sector work. The phys-ical separation of the worker from the workplace and the customer is now possible. These new forms of organisation and delivery bring with them new possibilities in the division of service labour, and for different types of skills from the past (Frenkel *et al.* 1999; Shire *et al.* 2002). Service work can be restructured, relocated and organised, as per a mass production model, with a high division of labour and continuous operations. Hence, refer-ences to manufacturing industrial systems with routinised production and rigid forms of control and labour subordination, as outlined by Braverman (1974), have been referred to in numerous call centre studies (Taylor and Bain 1999; Bain and Taylor 2000).

The explosion of academic research on call centres has matched the emergence and growth of the industry. For example, a search on 'Google Scholar' unearthed around 39,000 articles that mentioned call centres! In the social sciences, call centres embody many of the debates and discus-sions that resonate across disciplines. To begin with, there is the future of work in an age where the extensive application of ICT offers potential for new ways of doing things, expanding markets, developing new products and undertaking work in different ways (Burgess and Connell 2005). The very nature of, and even existence of, work has to be questioned in a context where many processes and functions can be automated or per-formed by customers. Technology brings with it the potential for better and new work, while also bringing the potential to eradicate jobs, deper-sonalise and deskill work (Castells 1996; Rifkin 1995; Reich 1991). Also evident are the processes of globalisation, whereby production and work become a temporary phenomenon in a spatial context. Coupled with the capabilities of ICT, the globalisation of service sector work means that new jobs can be located in new spaces and those regions and workers pre-viously excluded from the distribution of new wealth can be part of the global growth process. As Dicken (1998, p. 1) states 'something new is happening out there'. However, this process is also, paradoxically, a tem-poral process with regard to call centres since they can be located any-where, and can be superseded by new technological developments in service delivery. Production and employment can be shifted to new locations, supply chains can be extended and new networks and clusters of production can be generated (see the ICT and call centre cluster in Bangalore, India). For these reasons, these developments are not only about the transformation of work and production; but are also about the transformation of the workplace and the location of work (Dicken 1998).

The nature and development of call centres

Call centres are ICT-based workplaces that supply services to customers in diverse and remote locations via electronic media. Services provided include: inquiries, billing, marketing, advice, bookings, timetables, accounts and complaints handling. Functions that, at one time, would have occurred in-house and within divisions of large organisations (for example, accounts, bookings and marketing) can now be separated from head office and be relocated anywhere in the world, so long as the requisite ICT and labour requirements are met. In fact 'contact centre' frequently replaces the term call centre in the industry terminology, reflecting the fact that operations in addition to telephony, including email/web enabled transactions/texting, are now being utilised to deliver services to customers. The range of actual and potential services that can be located in a call centre is also increasing as communications technology is upgraded and new software is developed and applied.

Although each call centre has an ICT infrastructure platform that is supported by a workforce of operatives, managers and technical support staff, the appearance and configuration of each centre can differ (see Lindgren and Sederblad in this volume). There are differences between the types of ICT used, the services delivered and how production and employment are organised (Russell 2004). Nonetheless, the enduring feature of call centres is that they represent a new form of service delivery and employment. Work that was previously conducted within organisations is now performed elsewhere, often under conditions of mass production that, in turn, enables the realisation of economies of scale and an extended division of workforce labour.

In terms of industry evolution and development, it seems that call centres are moving into a phase of outsourcing and internationalisation (Taylor and Bain 2004). In Australia, it is estimated that approximately one-third of call centre functions are performed by specialist outsourced providers, and that this share continues to grow (Call Centres Net 2005). Processes of outsourcing and internationalisation offer cost and strategic opportunities and both are viable through ICT developments (see Srivastava and Theodore, and Taylor and Bain this volume). So, although examinations of the labour process within call centres have dominated earlier discourse on call centre operations (see Fernie and Metcalfe 1998; Taylor and Bain 1999), the evolution of call centres is now moving towards more automated processes and locations that are physically remote from head offices and customers.

While in the UK and the US the call centre industry has existed for more than a decade, in other countries it is almost a twenty-first century phenomenon (see for example, the studies of Greece and South Korea in this volume). Russell (2004) has likened the impact of call centres on service work and production to the impact of machinery on manufacturing work and production in the nineteenth century. New possibilities in delivering services are offered, and new forms of work are generated, often in new locations, requiring new skills and new forms of organisation (Frenkel *et al.* 1998). The type of work performed in call centres ranges from the mundane and routine (Bain and Taylor 2000), to skilled professional work (Colin-Jacques and Smith 2002). Work can be organised around strict factory regime type production models with tight monitoring and fixed production quotas (Bain and Taylor 2000) through to high performance work systems, built around teams and rewards (Kinnie *et al.* 2000). New job opportunities are offered in regions that have gone through post-industrial decline (Richardson *et al.* 2000) and new job possibilities and careers are offered to female workers (Hunt 2005). To 'new' service sector work we can add the strong attractor of internationalisation, or the more appealing term of globalisation. The call centre industry offers global possibilities and is growing in many less developed countries – leading to new jobs and a new international division of labour. This process is not, however, without its tensions and contradictions, with operatives sometimes being encouraged to hide their location and mask their ethnicity by pretending to be New Yorkers (tracking the progress of the Yankees), Glaswegians (Rangers or Celtic?) or Melbournians (comprehending Australian Rules Football) (see Taylor and Bain in this volume).

The call centre industry

Call centres are everywhere, except in the official industry data for each country. Even though they are more a way of conducting business via a different process, call centres can be labelled as an industry, operating in different locations and through different organisational forms than previously. The call centre industry is enigmatic, we know it is there, we can observe them and the activity taking place there, but it is difficult to collect 'official' call centre data. Call centres are an industry that has largely emerged over the past decade as a derivative of activities that previously occurred in other industries. In the call centre 'industry' elements of all other industries can be found. Indeed, call centres are a good example of how networking and business to business relationships can generate significant financial and efficiency gains. Likewise, all service occupations and professions can potentially be located in the call sector industry, including nursing (Colin-Jacques and Smith 2002) and social work (van den Broek 2003). While the origins of the industry can be traced to the restructuring of the delivery of front office customer relationships, the industry now encompasses both front and back office functions, as well as all service occupations.

As a result of the derivative nature of the industry, it is difficult to record accurately its scale and growth. So long as a service function can be removed from the need for the physical presence of a customer and the location of the industry service provider and relocated into a service centre, then call centre activity can be generated. As Russell states in his contribution:

> The quest for productivity improvement in the service sector, the availability of new information technologies and the growth of multi-dimensional forms of global competition has promoted the development of customer contact centres as a format for the rationalisation of info-service work. Thus, call centres usually entail a concentration of labour. As in the financial sector, although by no means restricted to it, branch offices and face-to-face encounters are replaced by over-the-phone, voice-to-voice encounters, which can be more tightly scripted, and controlled through standardised training in designated processes. Smaller, over-the-counter offices and service centres tend to give way to larger call centres in the process.

In some cases, work and production stay within the organisation, in others they are outsourced to specialist providers. These differences compound the problems of tracking the industry – as internal call centres are included in the activities of the home industry, the outsourced call centres are part of the communications sector of the economy.

The call centre industry tends to be dominated by banking, insurance, marketing, travel, telecommunications and public services (Burgess and Connell 2004). Since the industry is part of the ongoing restructuring and reorganisation of other industries, it is easy to lose sight of the fact that many call centre jobs and operations were previously located elsewhere. As a result, the high growth rates for the industry could be considered an illusion. This is because although the industry (through its consultants) reports high growth rates in terms of call centre jobs and sales, these are frequently based on activities that had previously occurred elsewhere in the economy. Thus, while claims of activity and job growth for the call centre sector are undeniable, it pays to be aware of the 'net' impact of call centre operations since there is a displacement effect (that is the substitution of call centre work for service work located elsewhere) and an efficiency effect (that is, fewer people doing similar work owing to the economies offered by ICT, outsourcing and an extended division of labour). This means that the 'spectacular' growth of the industry has to be put in the perspective of job losses taking place elsewhere in the economy.

This process of ongoing evolution and development of the industry has extended beyond national borders to include the externalisation or off-shoring of call centre activity to locations that offer extensive cost savings in service delivery. Hence, while there may be high rates recorded in terms of the number of centres and jobs being created in these offshore locations, again they represent the process of moving jobs that used to be in different locations, in different call centres or within host organisations.

Since the industry is difficult to classify and measure, much of the available 'information' on the industry originates from consultants and call centre service providers who have a vested interest in highlighting the growth and potential benefits of the industry to business, consumers, workers and communities. Consequently, there is a 'supply side hype' associated with projections and analysis, since extravagant claims regarding cost savings to businesses are made by those who have a vested interest in the ongoing expansion of the industry whether they be consultants, outsourcers or offshore service providers, or ICT system providers (see the chapter by Srivastava and Theodore in this volume).

Why the academic interest in call centres?

The apparent growth and proliferation in call centres has been matched by the level of research and publications on the topic. The call centre phenomenon embodies the 'mega' issues that impinge upon the future of work. As mentioned previously this includes: the impact of continued ICT development, the restructuring of organisations, the globalisation of business operations and the construction and delivery of service work. These are all profound issues that impact on work, work quality, jobs and living standards. As all of these developments are captured in the study of call centres, it not surprising that researchers have attempted to gain insight into these 'mega' issues. To date, there have been at least three collections that have brought together the contributions of call centre researchers from several countries, with differing points of focus. There is the examination of the reconstruction and delivery of service sector work (Frenkel *et al.* 1999), the labour process in call centres across the UK and Germany (Holtgrewe *et al.* 2002) and the issues associated with the organisation and management of the call centre workforce across several countries (Deery and Kinnie 2004).

Through call centres we can observe the contradictions and conundrums that are present in post-industrial work: the potential for subordination versus the potential for autonomy through ICT-based work (Russell 2004); servicing the customer versus being controlled by the customer (Korczynski 2001); developing new skills versus the erosion of traditional skills (van den Broek 2003); managing a workforce to deliver cost efficiency and service quality (Batt 2000); new work and careers for women versus a new and gendered division of labour (Belt 2004); and new jobs for depressed regions and economies versus a footloose industry (Burgess *et al.* 2005). As such, call centres represent manifestations of most contemporary debates concerning work, organisations and economic development.

The call centre industry is evolving, developing and going global. To date, the labour process has been central to call centre research, largely due to similarities with industrial production with respect to mass production and an extended division of labour, together with tendencies towards a routinisation of processes and product in conjunction with strict workforce controls (Bain and Taylor 2000). There are also questions surrounding employee motivation and commitment (Kinnie *et al.* 2000), the management of call centres to achieve inherently contradictory goals (Batt 2000) and forms of resistance and adaptation that occur in a context of extensive monitoring and surveillance (Barnes 2004). The issues of labour process, the organisation of work, and the control over work content and intensity remain important research issues. Then again, the enormity of the processes and implications of the call centre phenomenon are emerging as something that is profound in terms of the implications for work, employment and regulation.

In their review of the call centre labour process research, Holtgrewe and Kerst (2002) highlight the key issues as: a need to resolve the conundrums associated with standardisation versus customisation of services and operations; the ambiguity of customers as controlling and being controlled by the call centre process; the opportunities for new skills and the destruction of old skills; and the tension between the subordination of mass production versus the control over ITC service delivery. They state that 'call centre research has proved itself to be open to discoveries and able to learn from the field ... call centres are an organisational field in the process of construction' (Holtgrewe and Kerst 2002, p. 2).

Call centre research has emerged from many different perspectives, including management, sociology, economics, psychology, human geography and industrial relations. Call centres encompass challenges for existing analytical paradigms and, consequently, it is not surprising that the research is multi-disciplinary and exploratory, as in this volume. As call centres further evolve and develop, as the technology changes, then new issues and challenges will emerge for researchers. These issues do not stand alone, as there is an intersection between work organisation, the management of the workforce, the experiences of workers and the characteristics of the workforce.

In reporting on these issues, this book presents diversity with respect to the level or unit of analysis. Analysis focuses on micro-studies of specific call centres, industry studies of trends and developments across a number of call centres, meta-studies which place call centres within a broad context such as globalisation, and new service sector work. Specifically, this volume presents meta-analyses (Srivastava and Theodore; Taylor and

Bain); sector/country reviews (Lee and Kang; Weinkopf) and case studies
(Durbin; Houlihan; Rainnie and Drummond; Lindgren and Sederblad;
Russell). Different methodological approaches are also employed, with
examples of quantitative methods (Lee and Kang; Weinkopf) and qualitative methods including observation, interviews and focus groups (Russell;
Houlihan; Koskina). The volume also includes innovative participatory
research in the contributions from Russell, Houlihan, and Rainnie and
Drummond. A strength of this volume is that it showcases a diverse range
of approaches and research methods available for conducting call centre
research.

The organisation of the collection

The book is divided into sections on the basis of the level or breadth of
analysis employed. The following two Chapters, 2 and 3, encompass reflec-
tive overviews of the developments and trends in the industry, specifically
towards its relocation and internationalisation. The second grouping of
chapters presents studies of developments in national economies. The
third group examines specific research questions across a range of topics,
largely in the context of multiple case study research.

In Chapter 2 Srivastava and Theodore analyse the current offshoring
situation in the US using a meta-analysis of consulting white papers, industry trade reports and research studies to identify the trends and key issues
associated with offshore sourcing. They point out that 80 per cent of all
Chief Information Officers will be directed to move offshore at least part
of the technology services they provide to their businesses over the next
year. Although such moves are apparently demanded by investors anxious
to maintain shareholder value, and are strongly advocated by consulting
firms eager to sell services to cost-conscious managers, Srivastava and
Theodore maintain that there is a danger of backlash from workers and
consumers. To many, corporate gains from call centre offshoring have
come at the unacceptable price of deep domestic job losses. Consequently,
in this chapter they review: the principal motivations behind decisions to
offshore call centre services; the firms and sectors that have been at the
forefront; the services typically offshored; the new providers of offshored
services, emerging offshore markets and the constraints on offshoring,
before considering the implications of offshoring for US companies and
workers. The shift towards offshoring is supported by hyperbole and opti-
mism by consultants that masks some of the potential problems and limitations associated with offshoring, it is also reinforced by the pressure of
'shareholder value' and rating agencies that demand continuous cost
savings. This shift towards the external relocation of service sector work
destroys three myths often associated with service sector work – that it is
non-tradable and, therefore, faces natural protection from competition; it
is an area within organisations where there are few opportunities for

productivity gains; and it is in the realm of new 'service' work that comes with new skills and new careers (Reich 1991). Also, while offshoring at present tends to be centred on relatively low skilled and routine service functions, there is the potential to shift all service provision functions off-shore, including skilled and professional services.

In Chapter 3, Taylor and Bain continue the offshoring theme of Chapter 2 focusing on the situation in India where, since 2002, a large number of UK-based organisations have moved their call centre jobs. As with Srivastava and Theodore, they highlight the imperatives of cost savings, the exercise of greater controls of the labour process and increasing shareholder value as strong drivers of relocation decisions. They argue that decisions to relocate offshore appear to signify an unstoppable tide of activity, leading many to predict the extinction of the call centre industry in developed countries. This argument is based on the technologically deterministic assumption that call centres could be located anywhere, so long as the requisite technological infrastructure is in place. However, Taylor and Bain point out that one of the most important elements in the business case supporting moves to offshore is that, in moving voice services to India, companies have not only realised the potential for significant cost savings, but have also discovered a solution to persistent problems of labour utilisation and employee management (especially commitment and turnover). The appealing solution is to transfer call centre work processes from relatively high cost locations to relatively low cost locations. However, this process is not without its problems and contradictions. While there may be lower labour costs, there are other problems such as infrastructure reliability. Cost advantages have to be offset against customer resistance and dealing with the staff based in offshore locations, which raises questions regarding whether the labour processes and fundamental challenges for managers and workers are the same in offshore locations. Taylor and Bain provide a context that enables a better understanding of the complexities of work organisation and the employment relationship that are common to, and differ between, UK and Indian call centres. The imperatives are the same that drive and shape the process, but the context in which call centres are placed moderates their operations, management and forms of employee accommodation and resistance.

The next two chapters encompass relatively large scale surveys of call centres in Germany and Korea, respectively. In Chapter 4 Weinkopf focuses on work organisation and HRM strategies within 18 German call centres that differed considerably in the size, organisational structures and services provided. Specifically, Weinkopf set out to discover how German call centres were mastering the balancing act between cost efficiency and service quality in practice, and what implications these practices have on HRM strategies, job quality and employee job satisfaction. She found that the German call centres investigated were not, in every case, the 'sweat shops' or 'dark satanic mills' of the twenty-first century that have been described by other writers where workers are recklessly exploited (see Fernie and Metcalfe 1998; Knights and McCabe 1998). Moreover, the extent of monitoring and control was less comprehensive than in other countries such as the UK (Taylor and Bain 1999) – a finding endorsed by Lindgren and Sederblad in their chapter on call centres in Sweden. Weinkopf discusses potential explanations for such findings, concluding that the probability for improvement of working conditions in German call centres is optimistic, as the companies themselves are still seeking solutions to many of the issues raised in the research and she indicated that a trend towards higher-quality work practices could already be detected. Along with other chapters in the collection (Taylor and Bain; Koskina; Lindgren and Sederblad), this chapter highlights how national conditions and institutions, including social protection and social partnership, can shape the management of the labour process and affect the nature and quality of the jobs on offer.

Chapter 5 by Lee and Kang draws upon national survey data to provide an overview of the key features of employment relations and labour issues in South Korean call centres. Here, we have an industry that is less than five years old, with 60 per cent of operations being opened since 2000. The call centre survey was conducted in Korea between June and August, 2004 and is the first national survey of call centre operations and employment relations in Korea. It is also part of a global benchmarking survey of human resource practices and performance in call centres, which is co-managed by the Global Call Center Industry Project team comprising 15 countries, including the US, UK, Germany and Japan. Over 120 call centres were included in the Korean analysis. Key findings include the heavy dependence of outsourced providers and the contingent nature of the employment contract. Nearly 30 per cent of call centres were outsourced to subcontractors and over 90 per cent of customer service representatives (CSRs) in the remaining in-house call centres were non-regular (temporary) employees. The proliferation of call centres – whether in-house operations or subcontractors – has been led by management's strategic drive to enhance customer relations by using advanced information and communications technologies. Central operational dilemmas present in the organisation and management of call centres elsewhere are also present in South Korea – notably the dichotomy between service quantity and quality. Lee and Kang report that despite the enormous potential of the call centre sector for job creation (especially for the female workforce), to date the Korean government has made little effort to foster this sector, unlike some Western countries where the government is an active player in attracting call centre operations for the purposes of job creation in less favoured regions (Richardson, Belt and Marshall 2000).

The next two Chapters, 6 and 7, focus on call centre employees, their skills and their careers. In Chapter 6 Russell argues that a major issue in the analysis of the information-service work undertaken in call centres is just how closely such activity emulates or diverges from the industrial model of work and job design. In particular is it 'new' knowledge-based work, or is it routine service work? Russell questions what types of jobs *are* being generated and what type of jobs *can* be created for information-service workers. The discussion is largely confined to call centre development in Australia, where Russell set out to determine how agents perceive the skill demands of their work and to analyse factors that may influence such evaluations. The findings are based upon a workforce survey conducted within four call centres in the financial services, utilities, state licensing and transportation sectors. The survey was supplemented by extensive field observation. He found that, in the main, call centre work is routinised, repetitive and far removed from the information-service work idealisation. In this context, attempts to 'beef' up skill needs and empower workers run into problems of standardisation and control that are inherent to the construction of call centre work. It is workers who possess skills and credentials (in the case studies this group was mainly university students) that tend to be dissatisfied by call centre work and find the skill requirements to be mundane.

Chapter 7 by Durbin investigates issues relating to gender, skills and careers for women, where the central research question is 'do call centres offer women an opportunity to progress into management positions and if so, what types of management positions are they?' As Durbin points out, although call centres are overwhelmingly populated by women, senior management call centre teams are invariably, predominantly male. This chapter is based upon her fieldwork analysis of four call centres that form a part of two of the UK's largest financial services organisations. Durbin's findings offer a mixed picture of women's progression in call centres. There are opportunities for career progression available for women workers, but these are often associated with progression elsewhere in the organisation (away from the call centre) or progression within the industry, rather than within the call centre at which they are employed. Limits placed upon career progression within call centres include flat organisational structures, a lack of female role models in management positions, a lack of training opportunities, the male stereotyping of women's skills and attitudes towards a career, a lack of support and encouragement and family commitments.

Chapter 8 by Rainnie and Drummond is the only chapter in the book that examines the issue of union organisation in call centres. While the problems for unions aiming to organise in call centres have previously been reported (Taylor and Bain 2003), Rainnie and Drummond's account is different in that it provides a narrative of the organiser's experience at a non-metropolitan Australian call centre in the Latrobe Valley, 100 kilometres east of Melbourne. This chapter follows the progress of the organiser who led a successful campaign to unionise the call centre in question.

The model used was based on mobilising the community's exp‹
trade unions from an earlier period of industrial production in th‹
One of the strengths of this contribution is that it represents the organi‹
story as a participant in the process through action research, and reflects o‹
the application of the community-based model of union organisation (Lopez
2004). In this case, the success of the organising programme was supported
by reaching out beyond the workforce to the immediate community. In this
process, community barbecues were an important instrument in mobilising
support. The authors conclude that there are positive implications for
unions beyond the Latrobe Valley case. The movement of call centres from
city centres to small towns or offshore locations is being driven by labour
cost concerns, including the potential for union exclusion and more compliant workforces. The Latrobe Valley case study would seem to suggest that
the unionisation of call centres in remote and greenfield locations is possible. From several of the other studies (Taylor and Bain; Weinkopf;
Koskina) we can see that contextualisation is important, or in the case of
trade unions, an understanding of the historical and social context in which
call centres are located is important if they wish to recruit.

The next three Chapters, 9, 10 and 11, focus on managerial practices
and work organisation in call centres. In Chapter 9, Houlihan examines
the experiences of call centre managers – an area that has been relatively
under-researched to date. Data collection involved extended interviews
and observation within two call centres located in the UK. In the case of
the *Quotes Direct* call centre, Houlihan went one step further than Russell
(who sat alongside agents to conduct his research) undertaking full participation as a call centre agent over the period 1997–1999. In call centres, the
challenges and constraints that drove the daily practice of the managers is
highlighted. Houlihan contends that there are two competing dynamics
within call centre management. First, the variety, improvisation and innovations that prevail in managers' practices, that second, suggest the core
architecture of call centre organising (task routinisation, automated work-
flow control and intensive performance monitoring) and its underlying
tensions play a strong constraining role in the potential for call centre
managers to exercise their autonomy to manage. Thus, although call
centre management is typically portrayed as autonomous and deterministic, a picture emerges from Houlihan's research where, although managers
are required to be creative and innovative, they struggle with conflicting
role requirements and contextual constraints while frequently lacking the
support they need. The case studies also offer contrasts in the construction
of jobs and in the culture or expectations present at the workplace. Where
work is strongly monitored, controlled and targeted to strict performance
goals, call centre managers are constrained by the same processes and controls, often imposed by external managers. This study is important, since it
examines the experience of managers in call centres, highlighting how the
inherent tensions and paradoxes associated with call centre work are also

transmitted to call centre management. Just as call centre workers may push the rules and protocols as a way of accommodation and coping, it is evident that call centre managers may well engage in similar practices.

In Chapter 10 we move to Athens, Greece, where Koskina studies call centre work within four organisations. Specifically, Koskina investigates whether technology plays the most influential role in guiding the nature and management of call centre work. The underlying premise here is that call centre environments are an extension of 'Taylorist' approaches – where employees are perceived as a measurable entity with identifiable physical and mental traits that need to be manipulated by management to fit the requirements of production. Call centre workers, therefore, appear to occupy 'low-discretion' and 'low-trust' roles, in which management hold a high level of 'technical', 'detailed' and 'bureaucratic' control. Koskina's findings were completely at variance with such assumptions, as she found that the character of Greek call centre employment was markedly distant from the omniscient Anglo-Saxon Taylorist accounts reported by others. This confirmed that in none of the four organisational settings was the management and organisation of work driven by the use of technology. There was an idiosyncratic arrangement of call centre work that depended on trust and supported autonomy, mediated by cultural norms and expectations. National institutions and culture are important in determining how call centres are organised, how work is constructed and the responsibility and autonomy given to employees. Yet again, an understanding of context and differences between countries in call centre work and management is an emerging finding from the volume.

In the final Chapter, 11, Lindgren and Sederblad investigate the possibilities for worker autonomy in call centre work, introducing the concept of 'flexible autonomy' in relation to work organisation. Using observation, Lindgren and Sederblad studied three call centres based in Sweden, tracking one centre for five years. The rigid requirements of the labour process were moderated by providing employees with compensation in the form of breaks, amenities, private space, and in one case, relatively low call quotas. Context is important, as many of the Swedish call centres were unionised. Granting some autonomy was necessary as a means of ensuring labour retention and commitment. However, even within a more liberal context than their Anglo-Saxon counterparts, one of the call centres closed down following a protracted labour dispute.

Call centres as a work in progress

In summary, it is clear that the chapters contained in this volume add new perspectives and research approaches to the existing call centre literature. Moreover, it is evident that call centres will continue to evolve. They will become more globalised and service centre work will be further separated from the organisation and the customer. Further ICT developments will offer possibilities for greater automation, greater customer-directed service delivery and greater fluidity in call centre work and location. Consequently, call centres are definitely a work in progress, and the possibilities are potentially far reaching across the spectrum of service sector work. Service sector jobs can be reconfigured and relocated and even highly skilled service work has the potential to be delivered in different ways, through different mediums and in different locations. While cost advantages can be achieved through the relocation of service work, this is likely to continue. Although a number of these factors have been covered in this volume, there remain a number of key issues that warrant further research in the future. These are:

- The tracking of the call centre industry: This will continue to be a challenging task, particularly given the dearth of official data and the dependence for information on those who have a vested interest in the expansion of the industry.
- The offshoring debate: Call centres will continue to be relocated offshore in order to reduce costs and improve shareholder returns. The offshoring phenomenon raises many interesting issues. This includes a new international division of labour and jobs in poor regions of the globe. As Russell (chapter 6) points out, Australia, in common with other OECD countries, is now principally a service-oriented economy, with three-quarters of the workforce employed in a variety of business, personal and public services. Can the majority of jobs in the economy be relocated? What are the limitations to such a process? There will also be political and economic resistance to such moves from consumers, legislators, unions and the media. The reality of large scale transmigration of service sector jobs has yet to sink in, as we may be only witnessing the beginning of the process.
- Placing the call centre industry in context: This is also important in order to understand how national institutions, regulations and norms moderate the organisational architecture, working conditions and labour process that exists in call centres. Studies that compare and contrast call centres in different countries will be needed to fulfil this function. Even in 'low cost countries' the inherent nature of mass service production brings with it forms of accommodation and resistance to the tightly regulated regime of mass production.
- Public policy and call centres: Any analysis of public policy has been largely absent to date from call centre research. Issues include: skill generation and skill reduction, the role of subsidies and tax concessions in attracting call centres, the possibilities of careers and jobs in depressed regions, the development of operational codes of conduct (including OH&S) for the industry and the implications of delivering public services through call centres located in areas outside of the customer and taxpayer base.

- The design, management and delivery of call centre work: As the ICT possibilities and the range of services expand, then the issues surrounding delivery of quality services through the call centre platform will endure.
- The future of call centres: Now that call centres have had an operational record in advanced economies for over a decade, it will continue to be important to determine records in terms of attrition rates of centres and the sustainability of call centres and jobs in the face of technological and competitive pressures.

Other important issues covered in this volume that are worthy of future research include: the nature of call centre work and the skills required of such work, the quality of call centre jobs, career possibilities for women workers, and the challenges facing trade unions in membership recruitment.

Whether call centres are going to be a major feature in the future of work or just represent a temporary phase in the complete automation of many service work functions is debatable. This will depend, to some extent, on the speed with which interactive voice recognition occurs, and the speed with which web and email based transactions become more common – in either case this is likely to lead to a decline in the number of CSR positions.

Nonetheless, while call centres remain labour-intensive workplaces, the challenge for the future appears to be an achievement of balance between efficiency and quality in call centre management. Hence, while the growth in call centres continues to rise, in conjunction with the number of call centre workers, there will continue to be opportunities for research that explores how to balance the business case with employee needs in order to uncover 'best practice' models for call centre operations.

References

Bain, P. and Taylor, P. (2000) 'Entrapped by the "Electronic Panopticon"? Worker Resistance in the Call Centre', *New Technology, Work and Employment*, 15(1): 2–17.

Bain, P., Watson, A., Mulvey, G., Taylor, P. and Gall, G. (2002) 'Taylorism, Targets and the Pursuit of Quantity and Quality in Call Centre Management', *New Technology, Work and Employment*, 17(3): 170–185.

Barnes, A. (2004) 'Diaries, Dunnies and Discipline: Resistance and Accommodation to Monitoring in Call Centres', *Labour & Industry*, 14(3): 127–138.

Batt, R. (2000) 'Strategic Segmentation in Front-line Services: Matching Customers, Employees and Human Resource Systems', *The International Journal of Human Resource Management*, 11(3): 540–561.

Belt, V. (2004) '"A Female Ghetto?" Women's Careers in Telephone Call Centres', in S. Deery and N. Kinnie (eds) *Call Centres and Human Resource Management: a Cross-national Perspective*, Hants: Palgrave Macmillan.

Braverman, H. (1974) *Labour and Monopoly Capital: the Degradation of Work in the Twentieth Century*, New York: Monthly Labour Review Press.

Burgess, J. and Connell, J. (2004) 'Emerging Developments in Call Centre Research', *Labour and Industry*, 14(3), 1–13.

Burgess, J. and Connell, J. (2005) 'Sustainable Work: the Issues for Australia', in J. Goldney, B. Douglas and B. Furnass (eds) *In Search of Sustainability*, Melbourne: CSIRO Publishing, pp. 137–150.

Burgess, J., Connell, J. and Hannif, Z. (2005) 'Call Centre Development in the Australian Public Sector: Work and Service Delivery', Employment Studies Centre, University of Newcastle.

Burgess, J., Drinkwater, J. and Connell, J. (2005) 'Regional Call Centres: New Economy, New Work and Sustainable Regional Development', in A. Rainnie and M. Grobbelaar (eds) *The New Regionalism in Australia*, Ashgate: Aldershot, pp. 69–87.

Call Centres Net (2005) www.callcentres.net/, (accessed 3 August 2005).

Capelli, P., Bassi, L., Katz, H., Knoke, D., Osterman, P. and Unseem, M. (1997) *Change at Work*, Oxford: Oxford University Press.

Castells, M. (1996) *The Rise of the Network Society*, Oxford: Blackwell Publishers.

Colin-Jacques, C. and Smith, C. (2002) 'Nursing on Line: Experiences from England and Quebec', paper presented to the *International Labour Process Conference*, Glasgow, University of Strathclyde, April.

Datamonitor (2005a) 'Profiting from the North African Option', www.datamonitor.com (accessed 8 August 2005).

Datamonitor (2005b), 'The Future of Contact Centre Outsourcing in India and the Philippines, www.datamonitor.com (accessed 8 August 2005).

Deery, S. and Kinnie, N. (2004) *Call Centres and Human Resource Management: a Cross-national Perspective*, Hants: Palgrave Macmillan.

Dicken, P. (1998) *Global Shift*, 3rd edn, London: Paul Chapman.

Fairbrother, P., Paddon, M. and Teicher, J. (2002) *Privatisation and Globalisation of Labour*, Sydney: Federation Press.

Felstead, A., Jewson, N. and Walters, S. (2005) *Changing Places of Work*, London: Palgrave Macmillan.

Fernie, S. and Metcalfe, D. (1998) 'Not Hanging on the Telephone: Payment Systems in the New Sweatshops', Discussion Paper no. 390, Centre for Economic Performance, London School of Economics.

Frenkel, S., Tam, M., Korczynski, M. and Shire, K. (1998) 'Beyond Bureaucracy? Work Organisation in Call Centres', *The International Journal of Human Resource Management*, 9(6): 957–979.

Frenkel, S. J., Korczynski, M., Shire, K. M. and Tam, M. (1999) *On the Front Line: Organisation of Work in the Information Economy*, USA: Cornell University Press.

Holtgrewe, U. and Kerst, C. (2002) 'Researching Call Centres: Gathering Results and Theories', paper presented to *International Labour Process Conference*, Glasgow, University of Strathclyde, April.

Holtgrewe, U., Kerst, C. and Shire, K. (2002) *Re-organising Service Work: Call Centres in Germany and Britain*, Aldershot: Ashgate.

Hunt, V. (2005) 'Call Centre Work: Careers for Women?', *International Employment Relations Review*, 10(2): 111–132.

Kinnie, N., Hutchinson, S. and Purcell, J. (2000) ' "Fun and Surveillance": The

Paradox of High Commitment Management in Call Centres', *International Journal of Human Resource Management*, 5: 967–985.

Knights, D. and McCabe, D. (1998) 'What Happens When the Phone Goes Wild? Staff, Stress and Spaces for Escape in a BPR Telephone Banking Work Regime', *Journal of Management Studies*, 35: 165–194.

Korczynski, M. (2001) 'The Contradictions of Service Work: Call Centre as Customer-oriented Bureaucracy', in A. Sturdy, I. Grugulis and H. Willmott (eds) *Customer Service: Empowerment and Entrapment*, Hants: Palgrave.

Lopez, S. (2004) *Reorganising the Rust Belt*, Berkeley: University of California Press.

Reich, R. (1991) *The Work of Nations*, New York: Basic Books.

Richardson, R., Belt, V. and Marshall, N. (2000) 'Taking Calls to Newcastle', *Regional Studies*, 34(4): 357–376.

Rifkin, J. (1995) *The End of Work*, New York: Putnam.

Russell, B. (2004) 'Are All Call Centres the Same?', *Labour and Industry*, 14(3): 91–110.

Shire, K., Holtgrewe, U. and Kerst, C. (2002) 'Re-Organising Customer Service Work: An Introduction', in U. Holtgrewe, C. Kerst and K. Shire (eds) *Re-Organising Service Work. Call Centres in Germany and Britain*, Aldershot: Ashgate, pp. 1–16.

Tapscott, D. (1995) *Digital Economy*, New York: McGraw Hill.

Taylor, P. and Bain, P. (1999) 'An "Assembly Line in the Head": The Call Centre Labour Process', *Industrial Relations Journal*, 30(2): 101–117.

Taylor, P. and Bain, P. (2003) 'Call Centre Organising in Adversity', in G. Gall (ed.) *Union Organising*, London: Routledge, pp. 153–172.

Taylor, P. and Bain, P. (2004) 'Call Centre Offshoring to India: the Revenge of History', *Labour and Industry*, 14(3): 15–38.

van den Broek, D. (2003) 'Selling Human Services: Public Sector Rationalisation and the Call Centre Process', *Australian Bulletin of Labour*, 29(3): 236–253.

2 Offshoring call centres

The view from Wall Street

Snigdha Srivastava and Nik Theodore

Introduction

Offshore sourcing – the relocation of service jobs from the United States – has been widely described by business analysts as the 'next stage' of globalisation. A growing array of services formerly performed in the US, including customer support, claims processing and market research is now performed in a number of low-wage countries. By any measure, the practice of offshore sourcing is growing at a phenomenal rate. Forrester Research (2002) estimates that '(o)ver the next 15 years, 3.3 million US services industry jobs and $136 billion in wages will move offshore to countries such as India, Russia, China, and the Philippines'. This movement of jobs and wages is being led by business processes outsourcing (BPO), which includes customer interaction services such as call centres, help desks and technical support services.

This chapter provides a meta-analysis of consulting white papers, industry trade reports and research studies to identify the trends and key issues associated with offshore sourcing (also see Taylor and Bain this volume). Consultants and market analysts provide an important source of information regarding current offshoring practices and future trends. Consulting firms, in particular, provide the vision and expertise that allow companies to engage in offshoring, and they design many of the offshore sourcing arrangements that are being implemented by US firms. It is important to note, however, that although the reports reviewed here are touted by investors and embraced by senior executives, they should not be read as entirely objective statements about the pros and cons of offshoring. Many of the reports included in this chapter present best-case scenarios and are written in ways that endow offshoring with an aura of inevitability. As such the reports tend to highlight the short-run upsides of offshoring while remaining silent on longer-run concerns regarding skills formation, managerial oversight, employee morale and productivity, and the potential for consumer backlash. Consequently, they should be viewed as advocacy statements aimed at influencing corporate decision making. Nevertheless, they also form the foundation of the understanding of the business potential of offshoring, and are, in large part, responsible for how the issue is framed in the United States.

The first section of this chapter reviews the principal motivations behind decisions to offshore call centre services as described by outsourcing consultants and analysts, and identifies the firms and sectors that have been at the forefront of this business practice. This is followed by a review of the services typically offshored, the new providers of offshored services, emerging offshore markets and the constraints on offshoring. Finally, this chapter considers the implications of offshoring for US companies and workers.

The offshoring option

Offshore sourcing by US firms has been enabled by a set of factors that have made developing economies attractive and accessible locations for the relocation of customer service functions previously undertaken in the US. These include: (1) improvements in telecommunications capacity and reductions in telecommunications costs; (2) increased use of standardised enterprise software platforms that allow for a common set of employee skills across organisations; (3) widespread fluency in English (as well as Spanish, French and German) in parts of the developing world; and (4) marked pay differentials between US workers and workers possessing equivalent skills but who reside in low-wage countries (Accenture 2003; Auguste *et al.* 2002; McKinsey Global Institute 2003).

Although these four factors have aligned to create incentives and opportunities for offshoring on a large scale, the last factor – the existence of vast pay differentials between workers in different nations – is the chief motivation driving the decision to offshore services (see Agrawal *et al.* 2003; Riera *et al.* 2002; Shah 2003). The labour cost savings from offshore sourcing can be substantial. For example, a call centre agent in India is paid an annual salary between $2,400 and 4,000 per year, compared with an average annual salary of $16,000 to 20,000 in the United States (Steller 2003). Even when telecommunications and management costs are taken into account, call centre offshoring provides clear cost advantages to firms (see Table 2.1). The National Association of Software and Services Companies (Nasscom), an Indian trade association, estimates that US companies moving their call centre operations to India can expect average cost savings of 40 to 60 per cent. Parts of Eastern Europe, such as Romania, also offer cost savings of 40 to 50 per cent compared with US-based facilities. Operation costs are estimated to fall by 8 to 10 per cent even when call centres are exported to developed countries, such as Canada, Australia, Northern Ireland and New Zealand (Dawson 2003).

Although it has been a powerful inducement, reduction of labour costs is not the sole motivating factor in the call centre offshoring decision. In the past, intranational relocations of call centres allowed organisations to

Table 2.1 Cost of operating a call centre in Mumbai, India and Kansas City, USA 2002

	Amortized equipment cost ($/hour)	Other costs ($/hour)	Labour costs ($/hour)	Profit (20% markup in US, 100% in India)	Cost to client ($/hour)
Kansas City	0.25	0.14	10.00	2.08	12.47
Mumbai	0.35	0.21	1.50	2.06	4.12

Source: Dossani and Kenney (2003b).

pare labour costs by moving functions to low labour-cost areas, many of them rural (Dossani and Kenney 2003b; Atkinson 1995). However, faced with labour shortages, high attrition rates and poor education levels in lower-cost domestic locations, many US companies have opted for off-shore locations, where they have access to a large pool of heavily super-vised and well-educated workers to draw from (Dossani and Kenney 2003b). In India alone there is a large number of agents (estimates range from 40,000 to 100,000), mostly college educated, who handle calls and emails from customers abroad (Lachman 2004; Dawson 2003). Call centres around the world also invest heavily in workforce optimisation technolo-gies (Datamonitor 2004d), such as agent analytics and eLearning technolo-gies. Such investments attract large numbers of workers in developing nations who face chronic underemployment, creating a large pool of potential call centre workers. Low agent attrition rates are an added attraction in some of the favoured offshore locations. Agent attrition in outsourced call centres in the United States ranges between 50 and 100 per cent, compared with 10 per cent in the Philippines and 18 per cent in Ontario, Canada (Dawson 2003).

Who is offshoring?

'Offshoring' has quickly entered the business lexicon. It is important to note, however, that the term encompasses several distinct delivery models, the two most important being offshore *outsourcing* (when a company con-tracts out business process functions to an overseas service provider), and offshore *insourcing* (when a company sets up its own captive business process centres in overseas locations). Offshore insourcing is particularly popular with companies that wish to retain or maintain control over their offshore operations, which they do by operating wholly owned captive facili-ties in offshore locations while directly employing all workers in that facility. Companies typically have regarded their business processes and information technology capabilities as a form of service infrastructure that is necessary to support core, value-added activities. The costs of maintaining this infrastructure largely were seen as fixed. With the advent of offshore sourcing, companies are reconfiguring service delivery, making previously fixed costs variable, often by entering into arm's-length relationships with new service providers. In the case of offshore insourcing this relationship is established within the corporate structure of the offshoring firm (Celner *et al.* 2003). Captive centres have been seen as a way of maintaining control over service quality while also guarding against price increases that might be levied by offshore service providers over time. In contrast, offshore outsourcing to an independent entity offers its own advantages, in particular, the opportunity for rapid implementation. McKinsey and Co. (2003) estimates that revenues from captive offshoring are slightly more than double those of offshore outsourcing.

There are numerous examples of companies that have embraced global sourcing of call centre operations. Global customer service outsourcing firms, such as Convergys, APAC, Teleperformance and the ICT Group operate call centres in locations across the globe and offer varying price–skill combinations to clients. Others, such as America Online, have established captive call centres in low-wage countries to respond to their internal customer service problems (Steller 2003). Dell, Hewlett-Packard, Microsoft and other large IT companies have similarly launched captive call centre operations in India. Other US firms that prefer closer locations have started call centre operations in countries such as Canada and Mexico that yield labour cost savings along with the benefits of 'proximity to Americans in culture, legal system, infrastructure quality, trade relations and security as well as in geographic distance' (Dawson 2003, p. 5).

Services offshored

Offshore sourcing encompasses two broad subcategories: (1) offshoring of IT functions, and (2) offshoring of business processes (BPO), such as finance and accounting, customer services, human resources and payment services. Figure 2.1 categorises the types of services that are most commonly outsourced. At this time the BPO market is expected to grow faster than the IT offshoring market. The Aberdeen Group (2003a) reports that just 11 per cent of enterprises are outsourcing their software-related functions to offshore providers and only one-third of respondents to a global sourcing survey reported they expect to do so in the next five years. The offshore BPO market, on the other hand, is much wider (although many business processes involve a healthy dose of IT-related work) and is expected to experience strong growth. The McKinsey Global Institute (Agrawal *et al.* 2003) estimates that, in 2002, the BPO industry was valued at between $32 and $35 billion, or 1 per cent of the $3 trillion worth of business functions that could be performed remotely. Because major corporations have already realised significant benefits from offshore sourcing, BPO offshoring is forecasted to increase 30 to 40 per cent annually to 2007

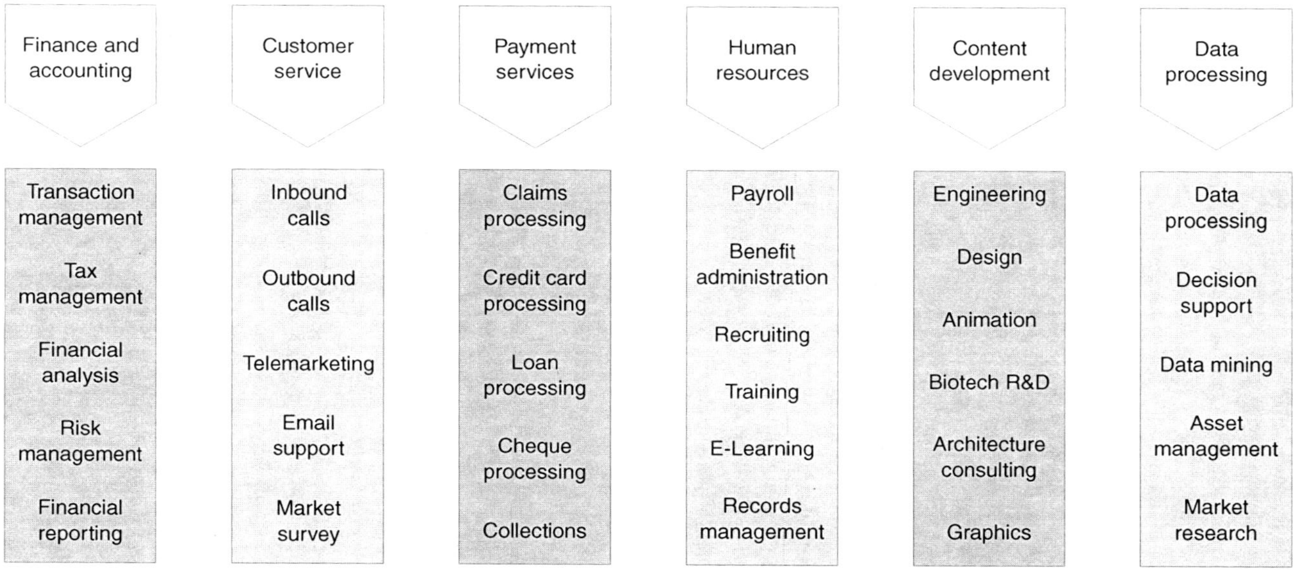

Finance and accounting	Customer service	Payment services	Human resources	Content development	Data processing
Transaction management	Inbound calls	Claims processing	Payroll	Engineering	Data processing
Tax management	Outbound calls	Credit card processing	Benefit administration	Design	Decision support
Financial analysis	Telemarketing	Loan processing	Recruiting	Animation	Data mining
Risk management	Email support	Cheque processing	Training	Biotech R&D	Asset management
Financial reporting	Market survey	Collections	E-Learning	Architecture consulting	Market research
			Records management	Graphics	

Figure 2.1 Most common services offshored (source: Shah 2003).

(Forrester Research estimates cited in Agrawal and Farrell 2003, p. 37). According to Gartner (2004b), in 2003, offshore BPO represented just 1 per cent of the total BPO market, but it is expected to increase to 14 per cent of the BPO market by 2008.

Call centres and customer contact centres, in particular, are expected to continue to be outsourced at a rapid pace. The Aberdeen Group (2003b) reported that 21 per cent of US executives (including 35 per cent of companies using call centres with 5,000 or more agents, and 33 per cent of companies with 500–1,000 call centre agents) plan increased use of outsourced contact centre services by 2006. Moreover, approximately 50 per cent of the executives surveyed indicated that they plan to outsource to a domestic provider, while 11 per cent plan to shift functions to Canada, 7.1 per cent to India and 5.3 per cent to the UK.

Key players in the offshore market space

The market for business process offshoring and call centre offshore service provision is a complex and fragmented one with multiple players operating in both overlapping and distinct niches. The following typology (adapted from Dossani and Kenney 2003b; Perkins 2003) classifies the players into the following categories.

MNC captives

Several prominent multinational corporations have used their existing operations in low-wage countries, such as India, to undertake business process activities that were previously conducted in other countries. Initially, many of the MNCs transferred business processes that were at the low-end of the value chain to their captive operations. Once they established a foothold, they ventured into more complex and business-critical tasks, such as establishing help desks and call centres. American Express, British Airways and General Electric were some of the first multinational corporations that set up business process operations in India. American Express set up its business process unit in 1993 in India, British Airways followed suit in 1996 and General Electric initiated operations in 1998 (Dossani and Kenney 2003b; Gupta 2002). More recently, AOL, Citigroup, Dell Computers, Hewlett Packard, HSBC, Oracle, Procter & Gamble and JP Morgan Chase have established call centres and other business process operation units in India, China and the Philippines.

Multinational outsourcers

Information technology consulting and outsourcing firms such as IBM, EDS, Hewlett-Packard and Accenture; call centre and customer relationship managers, such as Convergys, Sytel and Sykes; and payroll and accounting firms, such as ADP have entered the BPO and offshore call centre market. Many have incorporated offshore call centre services into their solutions portfolios and delivery capabilities. Most global service providers position themselves to deliver end-to-end, seamless service delivery augmented by industry consulting and other value-added project management expertise. These providers have established extensive networks of offshore call centres staffed by highly trained workers in low-wage countries. Typically, identifying the location for offshoring is at the discretion of the global service provider, which determines the most appropriate location for carrying out the activity (known as 'best shoring'). In response, these corporations have been operating service facilities in the Caribbean, Latin America, Ireland and Canada, as well as more recently in India, China and the Philippines (Datamonitor 2004c; Dossani and Kenney 2003b). In addition to their global reach, these large players have 'a fluent understanding of the market in their home countries as well as many other major markets: They have a proven track record in both IT services and business process outsourcing' (Perkins 2003, p. 1).

Home-shore anchored

Often started by émigrés with deep ties to their home countries, the owners and management teams of home-shore anchored providers are often based in the United States (or Europe), while the majority of technical staff and call centre agents are located in low-wage countries. Examples include Covansys, Cognizant and Syntel in the US and LogicaCMG in the UK. These firms understand the markets of the countries where they operate and have excellent delivery capabilities in specific niches (Perkins 2003). They offer their services at cheaper rates than global service providers, but do not offer as wide a range of services.

Established offshore

Many offshore firms were created to provide business process services to foreign firms. A number of these are venture capital-backed firms that were founded in the late 1990s to provide technical support, primarily via email and chat, to Internet ventures, such as Amazon.com and Yahoo! (Dossani and Kenney 2003b). After the bursting of the technology bubble in 2000, many of these service firms branched out into providing voice-based technical support to a broader clientele and some sought to secure accounts from multinational corporations. This market is highly dynamic, with a large number of entrants who face fierce competition, forcing survivors to 'pursue any business prospect' (Dossani and Kenney 2003b, p. 26). This often conflicts with their desire to maintain domain expertise and become industry specialists. One well-known survivor in the specialist market is Kale Consultants, a service provider based in Mumbai, that provides IT and business process services to the airline industry. 'Develop-ing domain expertise and becoming a specialist is difficult and risky, and yet, specialisation also offers the potential to occupy niches that may exist outside the ferocious competition found in the highly commoditised sectors' (Dossani and Kenney 2003b, p. 25).

Offshore IT subsidiaries

The best-known players in this category include large, India-based com-panies such as Infosys, Tata Consulting Services (TCS), Wipro, Satyam, HCL Technologies and Patni. Many started as 'body shops' but evolved to offer 'excellent global delivery capabilities within certain niches' (Perkins 2003, p. 1) and quality processes (Vijayan 2003). Their ability to secure contracts and maintain close customer relationships helped them branch out into new service areas such as software development, testing and maintenance services as well as BPO and call centre services. Large firms in this segment are maturing rapidly and expanding their service offerings to compete directly with global service providers (Perkins 2003). They also are building their internal staff capabilities (both in their home countries and increasingly in the United States) in business analysis, managerial experience and business process knowledge as they seek to expand their upstream IT consulting capabilities and BPO services (King 2003; Vijayan 2003).

Offshore non-IT subsidiaries

Many non-IT, established business organisations in offshore locations, such as India, have recently entered the offshore BPO market. Many of these firms, including Birlasoft and L&T Infotech, are subsidiaries of large, well-established business groups in their home country; they are well financed but have little or no prior experience in providing BPO services to foreign clients (Dossani and Kenney 2003b).

Emerging offshore markets

The vast majority of offshored business processes have migrated to coun-tries with three common features: low-wage rates for skilled, profes-sional positions; high educational attainment and workforce training levels; and the use of English as the main business language. Ireland and India have captured the largest shares of the business process offshoring market (Figure 2.2), and India is expected to increase greatly its offshore market share between 2005 and 2010. Canada and Israel, as well as Aus-tralia, South Africa and the Philippines, are also expected to receive increased US offshoring in the coming years (McKinsey Global Institute 2003).

Figure 2.2 Business process offshoring destinations (source: McKinsey Global Institute 2003).

US-based companies are especially likely to source call centre operations to Canada, the Caribbean, Latin America, India and the Philippines. According to Datamonitor (2004c and 2004e), out of a total of 2.86 million call centre agents in the US, 25,100 will be outsourced to Latin America by 2008, mainly from financial services, telecommunications and technology firms. From 2001 to 2007, the Caribbean and Latin American (CALA) call centre markets are expected to be the fastest growing in the world, increasing from 177,000 to nearly 700,000 agent positions by the end of 2007. Although the largest call centre market in the region is in Brazil, which currently comprises almost 50 per cent of all agent positions in the region, Mexico is forecasted to be the fastest growing market for outsourced positions in 2007, 'driven in large part by the efforts of outsourcers attempting to service the Spanish-speaking population in the US' (Datamonitor 2004c).

US-based firms will also increasingly rely on Canadian call centre agents (Datamonitor 2004b). In 2002 there were 450 outsourced call centres in Canada. This number is expected to rise to 600 by 2007. Correspondingly, the number of outsourced agent positions in Canada is projected to increase from 24,200 in 2002 to 36,800 by 2007 (Datamonitor 2004e).

Offshore sourcing: what are the implications for US workers and industry?

It was originally thought that business process offshoring would mainly affect back-office, highly routinised, low-end commodity work. However,

US companies are increasingly moving sophisticated, customer-facing, value-added jobs to overseas locations. According to industry observers, 'Any service activity that does not require direct physical contact is a candidate' for offshoring (Dossani and Kenney 2003a). The offshoring of service jobs began with the insourcing of low-end, routine software development and back-office operations by multinational corporations. The movement accelerated when companies outsourced their Y2K compliance work in response to a shortage of skilled workers trained in legacy systems. Next, companies tapped into their now well-established offshore connections to set up help desks and call centres dealing with various business processes in a variety of countries and regions, including India, Canada, Ireland, Central Europe and the Philippines.

The movement of white-collar jobs is affecting wage levels in the US. For example, Foote Partners, an IT advisory firm, found that offshoring IT-related functions has caused salaries in the sector to plummet, especially in the areas of application development and maintenance, call centres for technical support and some database work. According to Foote Partners, there is a high 'correlation between a decline in pay for skills and certification in areas that are actually moving offshore' (cited in Gonsalves 2004).

In addition to the impact on workers' wages caused by the migration of white-collar service jobs overseas, there are serious concerns regarding industry skills, expertise and competitiveness. Industry commentators have voiced concerns that as technical work moves offshore the 'experiential knowledge' that comes from years of solving technology problems will disappear, depleting the knowledge base and competencies of US industry.

> The reason [U.S. companies] can innovate without spending much on R&D is that they are learning each time they do an implementation. You build up that knowledge in those workers, and there's spillover as they move into other sectors, start new software companies or take a permanent job with a client. If we don't have that knowledge base here, we will lose out on that innovation and spillover.
> (Ron Hira IEEE-USA, quoted in Koch (2003). [Reprinted through the courtesy of CIO. Copyright © 2005 CXO Media Inc.])

Loss of experiential knowledge is one of several constraints that may limit the pursuit of call centre offshoring and quality of work is another (Pombriant 2004). Dell Computers, for instance, moved its call centre operations from an outsourced vendor in India back to the US in response to customer complaints regarding the problem-solving ability of Indian agents. Likewise, General Electric moved its appliance call centre work back to the US from India. It found that workers in India, who do not own many appliances, could not relate to customers' problems. Delta Airlines also returned outsourced services to the US from India because its customers complained of hard-to-understand accents. In recognition of these problems, Forrester Research predicts that offshore customer service and technical support centres will 'focus even more resources on accent neutralisation, English language proficiency, and both product and customer service training to eliminate service complaints by customers' (Forrester Research 2004).

Other limitations to the growth of call centre offshoring include:

Competition from speech-recognition software

Datamonitor (2004a) predicts that offshore call centres will face significant competition from speech-enabled, self-service technology, which offers a more cost-effective means to service low-level transactions. The research estimates that a firm may save 25 to 35 per cent of total costs by moving a US-based call centre to India, but if the same service can be performed by self-service technologies, an additional cost saving of between 15 and 25 per cent could be realised. There is evidence that corporations are spending more on speech-enabled technologies and this expenditure is expected to more than double to $1.2 billion in North America (McCue 2004).

Hidden costs

Executives often presume that offshoring (based on person-to-person comparisons) will yield cost savings of about 40 per cent or more. However, even though the wages of an equivalent worker in India may be 40 per cent less than in the US, there are transition and other hidden costs, including those related to vendor selection, staff turnover, contract management and vendor turnover which, when taken into account, yield smaller cost savings of 15–20 per cent in the first year (Davison 2003; Marks 2003). Moreover, the time and effort spent to transfer knowledge to the vendor is a cost rarely accounted for by the organisation. IT organisations, for example, 'experience a 20 per cent decline in productivity during the first year of an agreement, largely due to time spent transferring technical and business knowledge to the vendor' (Davison 2003).

High attrition rate

It has been widely reported that the single largest problem afflicting call centres worldwide is a rising staff turnover rate. Where India is concerned there have been recent reports that the attrition rate in the call centre industry now ranges from 20 to 50 per cent (Raman 2004; Merchant 2003). Although still lower than the 50 to 100 per cent attrition rate in the US, rising attrition rates inhibit the ability of call centres to grow swiftly over a short period of time in response to client demands. Employers are now

offering benefits, such as counselling, nutritional advice and higher education perks, in addition to extensive employee training, in an effort to ease
the high turnover rate, thus adding to their operating costs and constraining their comparative advantage and business growth.

Political backlash

The dislocation of manufacturing jobs is well documented and may be 'old
news', but the recent movement of white-collar call centre and customer
services jobs is generating widespread criticism amongst American consumers. Faced with intense political and public pressures, the State of New
Jersey directed eFund International, an Arizona-based customer service
company, to move the State's food stamp programme's electronic card
processing and customer service contract to New Jersey from Mumbai,
India. In response to the eFund fiasco, a bill passed the New Jersey legis-
lature that would require, among other provisions, all offshore call centre
calls to be routed to an American call centre if requested by the caller
(Dawson 2003). Meanwhile, legislation to restrict offshoring of call centre
work has been proposed or passed in various states, including Alabama,
Arizona, California, Connecticut, Idaho, Indiana, Louisiana, Maryland,
New York, Tennessee and West Virginia (Frank 2004). Such legislation
that makes effective export of data and call centres difficult could increasingly inhibit job migration in the future.

Conclusion

In most cases, the reports written by consulting firms and industry analysts
focus on the business advantages associated with offshoring – principally
the lowering of labour costs. These studies are largely silent on broader
questions regarding the implications of offshore outsourcing for firms,
workers and the US economy. Likewise, concerns regarding the long-run
implications for skills formation and industry expertise, for example,
receive little attention in these reports. Instead, management consultants
go to great lengths to insist that offshoring is an ideal route to firm competitiveness and a type of corporate 'best practice'. Furthermore, in trum-
peting the opportunity for US firms to pursue a course of radical cost
cutting by relocating production offshore, these reports tend towards a
portrayal of offshoring as an inevitable 'next step' in processes of globalisation. Perhaps this one-sided depiction of the advantages of offshoring
should be expected given that these consultants are aggressively selling
offshoring services to their corporate clients. But the reports reviewed in
this meta-analysis also implicitly raise more fundamental issues regarding
corporate restructuring and the role of US economic policy in aiding the
movement of service jobs offshore.

   The globalisation of services presents at least three challenges to the 'received wisdom' regarding the new service economy and its workforce. First, until recently, most customer-focused services were considered to be non-tradable, intrinsically local activities that had to be performed in close proximity to customers and end users. Advances in global telecommunications capacity and falling telecommunications costs are allowing companies to redraw the geography of service delivery. Second, senior executives previously saw few opportunities for productivity growth in companies' service infrastructure. Beyond trimming IT departments, outsourcing special projects and reducing headcount in 'non-core' job classifications, there were few productivity gains to be made. The opportunity to shift jobs to low-wage countries and cut wage bills has opened up new avenues for managers to pare labour costs. Third, in the wake of de-industrialisation and corporate downsizing, industry leaders and economic policy makers repeatedly have proclaimed that the future of the American workforce is in information technology and knowledge-based jobs, working as what Reich (1991) termed 'symbolic analysts'. Now these jobs are also subject to competition from low-wage economies – and the level of US job losses is projected to be staggering. 'You will see an explosion of work going overseas', says Forrester Research analyst John C. McCarthy (quoted in *Business Week* 2003, p. 52).

US firms are especially predisposed to pursue offshoring strategies. Wall Street profit pressures shorten businesses' time horizons to 90 days and punish those that fail to deliver continually improving expense-to-earnings ratios. A cost-cutting mentality has become embedded in a business culture that places a premium on low-cost, flexible staffing arrangements and on business models that deliver rapid results. In an environment that demands year-over-year, quarter-after-quarter growth, companies have turned to outsourcing as a way to reduce expenses and boost the bottom line. However, because domestic outsourcing agreements frequently fail to deliver anticipated results in terms of reductions in overheads and shifts in internal cost structures (Alvarez *et al.* 2003), firms increasingly are looking beyond the US market for the type of cost savings that could satisfy investor expectations. Given Wall Street's dominant role in shaping corporate objectives, it is not surprising that US firms account for approximately 70 per cent of the global offshoring market (McKinsey Global Institute 2003).

In order for US businesses to take full advantage of offshoring opportunities they must be free to reduce employment levels quickly in their US-based operations. This freedom requires that employment regulations provide workers only minimal employment safeguards. In experimenting with, and ultimately implementing, outsourcing strategies on a large scale, companies have been aided by business-friendly employment laws that allow conditions of 'at-will employment' to prevail in the US. These 'liberal employment and labor laws', according to the McKinsey Global Institute (2003, p. 2) 'allow companies greater flexibility in reassigning tasks and eliminating jobs. This flexibility is essential to capture offshoring opportunities effectively' since companies must be free to substitute relatively high-paid US workers with workers residing in low-wage countries.

Industry analysts and corporate executives often portray offshoring as simply an extension of the outsourcing practices that have been widely implemented by US firms. But unlike the classic make–buy decision that governs traditional outsourcing (e.g. evaluating whether it is more economical to produce versus purchase and distribute), offshoring has allowed US corporations to partake in labour-cost arbitrage. Global labour arbitrage arising from an imbalance between wage levels of qualified white-collar workers in different national markets allows companies to exploit this imbalance and profit from global uneven development. Global labour arbitrage has become especially attractive in current times of excess supply when firms are limited in leveraging product prices to their advantage. However, with labour costs constituting the bulk of production costs, it is not surprising that firms are leveraging the vast global wage differentials in response to the imperatives of cost control (Roach 2003). For instance, Roach (2003) indicates that in the US, worker compensations account for over 75 per cent of total domestic corporate income; given that wage rates in India and China account for 10 to 25 per cent of the prevailing wage rates for comparable workers, these wage differentials provide powerful incentives for firms to pursue global labour arbitrage. Companies experimenting with offshoring 'often start with labor-intensive processes that are neither critical nor core and expand only if they are happy with processing quality, reliability, and risk mitigation measures' (Riera *et al.* 2002, p. 10). This explains why early reports on offshoring focused on the outsourcing of low-end, routine, non-value-added tasks and functions – 'your mess for less' in the industry vernacular. From published reports it is clear, however, that low-end offshoring is just the beginning. As offshore providers ramp up their quality controls and improve customer service resolution rates, companies intent on cutting labour costs by offshoring services are seeking to move up the value chain to more complex – and integral – customer-facing functions. Gartner (2004a) estimates that the majority of offshore business process outsourcing is around customer-facing contact centres, including voice, email, and chat and the remainder is in non-customer-facing, internal support functions, such as accounting, human resources, legal and procurement.

Attracted by the promise of slashing payroll costs, and the maturing of offshore platforms and service delivery, leading corporations have turned to call centre offshoring as a way to bolster the bottom line quickly while at the same time sending a signal to customers and investors that they are serious about cost control. According to *Computerworld* (King 2003), this year, 80 per cent of 'all CIOs will have direct marching orders to move offshore at least part of the technology services they provide to their businesses'. Although seemingly demanded by investors anxious about maintaining shareholder value and strongly advocated by consulting firms eager to sell services to cost-conscious managers, there is a danger of a backlash by workers and consumers. To many, corporate gains from call centre offshoring have come at the unacceptable price of deep domestic job losses.

Acknowledgements

We thank Rich Feldman, Marcus Courtney, Marc Doussard and Jamie Peck for reviewing earlier drafts of this report. Cedric Williams assisted with the figures used in this report. Finally, we would like to acknowledge the financial support of the Ford Foundation.

References

Aberdeen Group (2003a) 'Global Sourcing Benchmark Report: Balancing Supply Cost, Performance, Risks in an Uncertain Economy', Press release, Aberdeen Group, Inc. (July).

Aberdeen Group (2003b) 'Looking for Call Center Clout: Look Outside', Press release, Aberdeen Group, Inc. (September).

Accenture (2003) 'Control in the Manufacturing and Consumer Industries – Research Summary', available at www.accenture.com/xdoc/en/services/hpb/insights/outsourcing.pdf (accessed 13 December 2004).

Agrawal, V. and Farrell, D. (2003) 'Who Wins in Offshoring', *McKinsey Quarterly* (Special Edition: Global Directions): 37–41.

Agrawal, V., Farrell, D. and Remes, J. A. (2003) *Offshoring and Beyond*, San Francisco: McKinsey and Company.

Alvarez, E., Couto V., Disher C. and Gupta A. (2003) 'Business Process Outsourcing and Offshoring: Proceed ... But With Eyes Wide Open'. Consultants' briefing by Booz, Allen, Hamilton at *cio.com* (22 October – date of publication): www2.cio.com/consultant/report1858.html (accessed 29 November 2004).

Atkinson, R. (1995) 'Technology and the Future of Metropolitan Economies', *Assessing the Midwest Economy*, Federal Reserve Bank of Chicago: MA-4.

Auguste, B. G., Hao, Y., Singer, M. and Wiegand, M. (2002) 'The Other Side of Outsourcing', *McKinsey Quarterly*, 1: 53–63.

Business Week (2003) 'Is Your Job Next?' (3 February).

Celner, A., Gentle C., Lowes P. and Nikolis P. (2003) *The Offshoring Imperative: Managing the Forces in the New Mandatory World of Offshoring*, Deloitte Consulting.

Datamonitor (2004a) 'Offshore Call Center Agents to Compete with Speech-Enabled Self Service Technology', Press release, Datamonitor, plc (18 November – date of press release): www.datamonitor.com/~3222444616ce42da8895793b1513828c~//home/press/article/?pid=031CD494-C679-4F1D-8864-B5BFB1FD47E4&type=PressRelease (accessed 7 December 2004).

Datamonitor (2004b) '3,000 Call Centers to be Extinct by 2008', Press release, Datamonitor, plc (27 August – date of press release): www.datamonitor.com/~3222444616ce42da8895793b1513828c~//home/press/article/?pid=9DD71171-BB55-4AB5-8FD4-16344078137F&type=PressRelease (accessed 7 December 2004).

Datamonitor (2004c) 'Fear of Job losses from U.S. Call Center Outsourcing to Latin America is Much Ado About Nothing', Press release, Datamonitor, plc (25 May – date of press release): www.datamonitor.com/~3222444616 ce42da8895793b1513828c~//home/press/article/?pid=9C1D5ACA-6F61-4250-9DA8-1E18EE8FBDB9&type=PressRelease (accessed 7 December 2004).

Datamonitor (2004d) 'Call Centers to Lavish over $1bn on Improving Customer Service, Agent Performance and Overall Efficiency', Press release, Datamonitor, plc (27 February – date of press release): www.datamonitor. com/~3222444616ce42da8895793b1513828c~//home/press/article/?pid=82F8888F -95C2-4CB4-8154-286C8C806F76&type=PressRelease (accessed 7 December 2004).

Datamonitor (2004e) 'Offshore Call Center Outsourcing: What Threat to Jobs?', Press release, Datamonitor, plc (23 January – date of press release): (Press release no longer available on Datamonitor's web site).

Davison, D. (2003) *Top 10 Risks of Offshore Outsourcing*, Stamford: Meta Group Research (14 November). Reprint available at http://searchcio.techtarget.com/ originalContent/0,289142,sid19_gci950602,00.html (accessed 11 December 2004).

Dawson, K. (2003) *International Outsourcing*, New York: Call Center Magazine, CMP Media, Inc. (August): www.nxtbook.net/live/CMP/outsourcesupplement/ (accessed 23 November 2004).

Dossani, R. and Kenney, M. (2003a) 'Lift and Shift? Offshoring Service Provision to India', A Power Point presentation at the Information Policy Institute's symposium titled *Offshoring & Outsourcing: What's in America's Best Interest?*, Washington, D.C., 11 December 2003.

Dossani, R. and Kenney, M. (2003b) 'Went for Cost, Stayed for Quality? Moving the Back Office to India', Stanford: Asia-Pacific Research Center, Stanford University (November – publication date): http://aparc.stanford.edu/publications/20337/ (accessed 3 November 2004).

Forrester Research (2002) '3.3 Million U.S. Services Jobs to go Offshore', Press release, Forrester Research (5 December – date of press release): Press release unavailable online.

Forrester Research (2004) 'Forrester Releases Top Technology Trends for 2004', Press release, Forrester Research (7 January – date of press release): http://www.forrester.com/ER/Press/Release/0,1769,874,00.html (accessed 29 November 2004).

Frank, D. (2004) 'The State of Offshoring', *Federal Computer Week* (14 June – date of publication): http://www.fcw.com/fcw/articles/2004/0614/feat-global2-06-14-04.asp (accessed 3 November 2004).

Gartner (2004a) 'Gartner Says Offshore BPO Industry to grow 65 per cent in 2004', a Gartner press release (18 May – date of press release): www4.gartner.com/press_releases/asset_79327_11.html (accessed 29 November 2004).

Gartner (2004b) 'Gartner Says Business Process Outsourcing in Europe will reach €25 billion in 2004, a Gartner press release (3 March – date of press release): www4.gartner.com/press_releases/asset_61528_11.html (accessed 29 November 2004).

Gonsalves, A. (2004) 'Outsourcing is Key as IT Salaries Spiral Downward', *InformationWeek.com* (14 January): www.informationweek.com/showArticle.jhtml? articleID=17300898 (accessed 15 January 2004).

Gupta, S. (2002) 'Demystifying Offshore Outsourcing', *CMA Management Magazine* (November): www.managementmag.com/index.cfm/ci_id/1847/la_id/ 1.htm (accessed 13 December 2004).

King, J. (2003) 'IT's Global Itinerary: Offshore Outsourcing Is Inevitable', *ComputerWorld* (15 September – date of publication): www.computerworld. com/managementtopics/outsourcing/story/0,10801,84861,00.html (accessed 13 December 2004).

Koch, C. (2003) 'Offshore Outsourcing: The Politics', *CIO.com* (1 September – date of publication): www.cio.com/archive/090103/backlash.html (accessed 11 December 2004).

Lachman, M. L. (2004) *The New Exports: Office Jobs*, New York: The Urban Land Institute and The Paul Milstein Center for Real Estate, Columbia Business School.

Marks, S. (2003) 'Offshore Outsourcing: Business Boon or Bust?', *NetworkWorld* (22 December): www.nwfusion.com/power/2003/1222offshore.html (accessed 24 November 2004).

McCue, A. (2004) 'Offshoring Faces Automated Call Centre Threat', *Silicon.com* (22 November): www.silicon.com/research/specialreports/offshoring/ 0,3800003026,39126052,00.htm (accessed 23 November 2004).

McKinsey & Co. (2003) *New Horizons: Multinational Company Investment in Developing Countries*, San Francisco: McKinsey & Co.

McKinsey Global Institute (2003) *Offshoring: Is It a Win-Win Game?*, San Francisco: McKinsey & Company.

Merchant, K. (2003) 'Why BPO is the Next Big Thing: Call Centres', *FinancialTimes.com* (5 February): Reprint available at www.nasscom.org/artdisplay.asp? Art_id=1619 (accessed 15 November 2004).

Perkins, B. (2003) 'Selecting the Right Offshore Vehicle', *Computerworld* (15 September): www.computerworld.com/managementtopics/outsourcing/story/ 0,10801,84826,00.html (accessed 13 November 2004).

Pombriant, D. (2004) 'On and Offshore: Call Center Economics', *ECT News Network* (15 September): http://www.crmbuyer.com/story/36600.html (accessed 11 December 2004).

Raman, S. (2004) 'Greying of India's Call Centres', *BBC News* (5 August).

Reich, R. (1991) *The Work of Nations*, New York: Alfred A. Knopf.

Riera, A., Sinha, J. and Shah, A. (2002) *Passage to India: The Rewards of Remote Business Processing*, Boston: Boston Consulting Group.

Roach, S. (2003) 'The Global Labor Arbitrage', Morgan Stanley's *Global Economic Forum: Latest Views of Morgan Stanley Economists* (6 October): www.morganstanley.com/GEFdata/digests/20031006-mon.html (accessed 15 February 2005).

Shah, R. (2003) *Who Moved My Job? A Cost/Benefit Analysis of the Impact of Offshore Outsourcing on Employment and Other Demographic Constituents*, India: New Horizons Consulting.

Steller, T. (2003) '3 Call Centers Take Jobs to India', *Arizona Daily Star* (14 May): www.azstarnet.com/ (accessed 11 December 2004).

Vijayan, J. (2003) 'India Inc., Still Going Strong', *Computerworld* (15 September): www.computerworld.com/managementtopics/outsourcing/story/0,10801,84834,00 .html (accessed 13 November 2004).

3 Work organisation and employee relations in Indian call centres

Phillip Taylor and Peter Bain

Introduction

Just as academic studies had come to terms with the development of the call centre as a distinctive form of work organisation within developed countries, the emergence of a globalised dimension to the location of this telemediated customer servicing function presented fresh challenges to the research community. Over the last few years the extent of offshoring – to use the term which increasingly entered business parlance in the US in the 1990s – of voice services to India, by companies based in the English-speaking geographies, has accelerated (see Srivastava and Theodore, this volume). For example, since 2002 the roll call of UK-based organisations to have migrated call centre jobs includes Prudential Assurance, Aviva, HSBC (in three separate tranches), Lloyds-TSB, British Telecom, National Rail Enquiries, Reality and Client Logic. These decisions seemed to signify an unstoppable tide of offshoring, leading many to predict the extinction of the call centre industry in the UK. Sensationalist media speculation was often based on the technologically deterministic assumption that call centres could be located anywhere, so long as the requisite infrastructure was in place (see for example the *London Evening Standard* 11 September 2003).

In addition, it was believed that India's attraction as a desirable destination rested on its low cost base *and* its specific country advantage, its deep reservoir of literate, fluent English-speakers. Statements by corporate decision-makers reinforced the perception that what distinguished India from other low-cost offshored locations were its unique workforce characteristics. The Chief Executive of HSBC, Sir Keith Whitson, achieved notoriety when he compared the company's UK workers unfavourably with their Indian counterparts, whom he lauded for their hard work, appearance and quality of work (*Financial Times*, 6 August 2002). In similar vein, Susan Rice, Chief Executive in Scotland of Lloyds TSB stated:

> Offshoring could be the best thing that ever happened to financial services in Scotland. Why? Because most of it is going to India where you

get graduates. You get a different type of person going into call centres, making it their career. It is a different beast, the call centre or the contact centre there.

(*Sunday Herald*, 18 April 2004)

Thus, an important element in the business case in support of offshoring was, and remains, the contention that in migrating voice services to India, companies have not only realised the potential for significant cost savings, but have also discovered a solution to the persistent problems of labour utilisation and management, often associated with call centre work organisation in the UK, North America and Australia. It suggests that the work processes of the call centre can be transferred from the developed to the developing world without difficulty.

However, contrary to this optimistic perspective, we maintain that work organisation in Indian call centres tends to replicate, if not exacerbate, many of the difficulties experienced in the developed world. In order to understand fully why this is so, it is necessary to take account, however briefly, of the dynamics of the Indian industry's growth, and how it was the product of Western corporate strategies, which were themselves shaped by more general political and economic forces and influences. Essentially, Indian call centres, serving English-speaking markets, emerged as an integral part of a wider business processing (BPO) sector, dependent on decisions taken in boardrooms in London, New York and elsewhere to offshore, initially at least, the most standardised and least risk-laden processes. Further, it seems apparent that the English-speaking Indian agents experience a series of contradictions in their relationships with their Indian employers and managers, which are intimately connected to the nature of the servicing of Western clients and customers. In the final analysis, the growing integration of the world economy, to the extent that this has occurred, provides the broadest context for these transnational service flows.

In this chapter specific consideration will be given, first, to the recruitment, selection and training of Indian agents, acknowledging the importance of cultural and linguistic issues. Second, the nature of the labour process and task performance will be examined with reference to the ways in which the interests of Western companies, and their customers' requirements, create distinctive Indian call centre workflows and methods of control. Third, the consequences of these work patterns for the Indian workforce will be assessed. Fourth, forms of workplace governance will be explored with particular emphasis placed upon employee representation and the possibilities for collective organisation and trade unionism. Throughout, we engage with the theoretical issues, debates and empirical material from recent academic literature. Accordingly, this will provide a framework for understanding the features of work organisation and the employment relationship that are common to, or differ between, the UK and India.

From a three-year-long project investigating the extent and nature of call centre offshoring to India we draw upon data from three sources; first, a complete audit of the Scottish call centre sector conducted between February and July 2003, and second, fieldwork carried out in India between January and June 2003. Combined with documentary evidence, both sets of findings were incorporated in a published report.[1] Additional evidence of trade union developments comes from extensive interviews with officers of Indian and UK trade unions and from national and international conferences on offshoring.

Globalisation and UK and Indian call centres

In the 1990s, corporate strategies increasingly focused upon expanding 'globalised' operations. Indeed, this was the decade of globalisation, in the sense that the widespread acceptance of the term signified the emergence of a new orthodoxy in the way the international capitalist economy was conceived. For many, dubbed 'hyperglobalists' by Held *et al.* (1999), the world had become dominated by an unrestrained capitalism, where manufacturing, and later, the service industries, sought out low cost havens in the developing world. This, it was claimed, produced a 'genuinely borderless' global economy which represented the 'death of distance' (Cairncross 1997; Ohmae 1995). Of course, these overblown assertions have been successfully challenged by theoretical and empirical challenges, not least in relation to the continued importance of the role of the state, the unevenness of capitalist development and the dominance of the powerful nations (Callinicos 2001; Harman 1996; Hirst and Thompson 1996; Ruigrok and van Tulder 1995). A recent study confirms that 'over the past two decades [foreign direct investment] FDI flows have remained concentrated within the developed world', which receives 60–80 per cent of all inflows (Kohler 2003, p. 25). Thus, international trade and economic activity remain substantially within and between the 'triad' of North America, Western Europe and Eastern Asia (Dicken 2003). Whilst recognising the transformative role played by transnational corporations (TNCs) in extending the research of capital accumulation, it is also true that they are not 'genuinely footloose', but are largely locked into these flows of trade and investment, retaining close relations with their identifiable home states.

Nevertheless, despite the persistence of countervailing forces and the necessary acknowledgement that globalisation represents sets of tendencies rather than a single, predetermined trajectory with inevitable outcomes, the growing interconnectedness of the world economy is undeniable (Dicken 2003; Held *et al.* 1999). One does not need to accept the tendentiousness of Castells' (2000) paradigmatic reconfiguration of the global economy – the network society – to acknowledge the potential for transformation set in train by the growth in sophistication and utilisation of computer power and telecommunications capacity. In particular, it is the developments in information and communication technologies (ICTs) which have facilitated or enabled, but not determined, the dispersal and relocation of office-based service work (Dicken 2003, p. 85). Consequently, notwithstanding necessary caveats, the remarkable growth in the globalisation of service work, representing the emergence of a new international division of labour, is set to continue.

In order to understand the specific place that the call centre occupies within these broader trends to offshore and/or outsource services, it is necessary to reflect upon its origins, emergence and subsequent rapid growth. From inception, the innovative organisational form of the call centre provided opportunities for companies to achieve considerable cost savings, and profit generation, through the centralisation of previously localised servicing and sales activities (Marshall and Richardson 1996). Economies of scale, overhead reductions and new selling opportunities were facilitated by the distinctive integration of telephonic and computer technologies, notably the ACD system, which enabled remote customers to be connected in real time to servicing centres, and were realised through novel forms of labour utilisation and work organisation (Bain *et al.* 2002; Callaghan and Thompson 2001; Taylor and Bain 1999; Taylor *et al.* 2002).

Thus, it is commonly acknowledged in the academic literature that call centres are constitutive of two competing logics. For Korczynski (2002) this is the simultaneous requirement to be cost-efficient and customer-oriented, a formulation which has been expressed, alternatively, as the perpetual contradiction between the competing managerial priorities of quantity and quality (Bain *et al.* 2002; Taylor and Bain 1999, 2001a). However, behind this apparent consensus lie important differences of interpretation. Frenkel *et al.* (1999) stress the customer-orientation element, believing that the need for customised responses, even though these may be delivered in high volumes, places strict limits on the degree of standardisation and routinisation, rendering restrictive management control strategies redundant. In this upbeat version the intangible quality of customer interaction inhibits 'management's ability to measure outputs' (Frenkel *et al.* 1999, p. 139), and call centre agents, who increasingly resemble creative, knowledge workers, are given 'expanded levels of discretion' (Frenkel *et al.* 1999, p. 69).

The case against this optimistic account, which emphasises the cost-reduction, or profit-maximising logic is compelling. It is recognised in the literature that while there are differing levels of heterogeneity in call centres, the routinised, Taylorised 'production-line call centres proliferate' (Bain *et al.* 2002, p. 184; Batt and Moynihan 2002, p. 14; Houlihan 2002, p. 98; Thompson *et al.* 2004, p. 148; Taylor and Bain, 2001a, p. 48; Taylor *et al.* 2002, p. 147). This is evident in relation to Batt and Moynihan's (2002) three models (mass production, mass customisation or hybrid), Taylor and Bain's (2001a) quantity/quality spectrum, Hutchinson *et al.*'s (2000) distinction between tightly controlled 'transactional' work and 'relational' customer interaction and in Houlihan's (2002) convincing argument that variation exists within a common low-discretion, high-commitment regime. Consequently, there is an inescapable tension between the imperative to maximise call volumes and the need to provide excellent customer service, which is reflected in qualitative and quantitative monitoring and the imposition of performance targets of all kinds (Bain *et al.* 2002). However, in the final analysis, quality considerations, for all but the highest value customers and/or the most complex services, tend to be subordinated to the dominant need for efficiencies and call volumes. Customer service, ultimately, is driven not by some autonomous logic, but by the overriding requirement that it delivers profitable outcomes. The very rationale of the call centre lies in its promise to cut costs and raise profits.

Given this central objective, there was always, crucially, a geographical dimension to the emergence of the call centre. Profit-maximising strategies depended upon the physical relocation of facilities within the developed countries to cheaper locations – particularly in relation to labour and property costs – where skilled labour was available and regional economic policy was conducive (Bain and Taylor 2002a; Bristow *et al.* 2000; Richardson and Belt 2001; Taylor and Bain 1997, 2001b, 2003). These trends became most evident in the distinctive regional and urban clusters that emerged in the developed countries. In one sense, then, corporate decisions to situate call centres in India can be regarded as an extension, however radical in geographical reach, of the means by which cost reduction and profit maximisation in customer servicing have been realised through organisational restructuring and spatial relocation. Yet, despite the availability of 'distance shrinking' technologies, and the potential for call centres to be located in the developing world, until recently they remained concentrated in the home-base countries of customers. However, it is possible to identify the principal reasons for the transformation of offshoring potential into globalised reality.

The successful experience of early offshorers to India (American Express, General Electric and British Airways) from the mid-1990s onwards proved influential.[2] While initially these were not interactive call centre operations, but in-house, directly owned back-office processing centres, 'captives' in Indian industry terminology, they demonstrated the capabilities of both the Indian workforce and the improving telecommunications structure. Tightening labour markets in the US in the 1990s increased the flows of offshored IT and back-office work, and led to GE opening their first Indian 'voice' operation, which served as a model for the burgeoning 'third party' call centre sector. Clearly, migration has also been facilitated by the success of the longer established Indian software industry, and the Indian government's commitment to providing a supportive business environment, as part of its central objective in attracting foreign investment.

Undeniably, though, the principal factor driving call centre migration has been the promise of overall cost savings of 40–60 per cent (Nasscom

2002, p. 12), realised through India's outstanding country advantage and the low labour costs of its English-speaking graduate workforce estimated 'at 70–80 per cent for offshorable processes' (Nasscom 2003, p. 65). For organisations in the English-speaking geographies of the US, the UK and Australia, the compulsion to reduce the cost base of their call centre operations intensified in the wake of the collapse of the speculative boom of the late-1990s (Brenner 2002), which dislocated the 'new economy' of technology, media and telecommunications (Fransman 2002; Henwood 2003). Given widespread recessionary pressures and acute problems in particular sectors, offshoring seemed to offer a 'spatial fix', to apply Harvey's general concept (1982, 2000), to the problems of profitability and accumulation facing those companies engaged in servicing activities.

Although Indian call centre employment has grown – and will continue to grow – rapidly, a sense of proportion is necessary. Distilling all available evidence, we calculated that, at June 2003, between 75,000 and 115,000 were employed in Indian call centres (Taylor and Bain 2003, p. 84), a total that by December 2004 had risen by around 60 per cent, but no more than 20,000 were UK-facing call centre jobs. Moreover, Vanaik (2004, p. 54) calculated that, at 2002, call centre employment in India amounted to a mere 0.002 per cent of the country's aggregate workforce.

Indian call centre workflows

Survey evidence of UK companies confirms that the types of call centre work being migrated to India are overwhelmingly the most routinised. They report migrating the following workflows: overflow and out-of-hours calls, basic customer services, processed stages of insurance claims, directory enquiries, ticket fulfilment, debt recovery/reminders/collections, telemarketing/telesales and technical support/IT help desks (Taylor and Bain, 2003, pp. 53–54). With the exception of the last more complex work, the majority consist of the most standardised processes, most of which derive from the financial services sector. An interesting statistical finding from Scotland (Taylor and Bain 2003, p. 62) is that those companies who have offshored operate call centres whose mean workforce size is more than double that of the sector average. What is significant here is that the larger the firms, with the most sizeable establishments, the greater the likelihood that the call centre will achieve the higher call volumes of a transactional, routinised kind which lend themselves to offshoring.

Most often, offshoring involves not the wholesale substitution of call centre services in India for those in the UK, but rather companies engaging in organisation-wide restructuring of processes, which leads to 'slicing off' the simplest, standardised and least risk-laden workflows for migration, like a form of exported Taylorism. In the process, the same cost-reduction logic that created, grew and continues to drive the call centre in the advanced countries is now fuelling offshoring to India. Consequently, it is producing forms of work organisation which lie at the extreme of the quantitative end of the spectrum of call centre work.

Viewed from the Indian perspective, its outsourced call centre sector has developed as a niche market for the developed nations' call centre industry, where 'the model today for a third party player can best be described as an offshore based replication of on-site, or onshore activities' (Venkatraman speech, 12 June 2003). Frequent claims are made that the Indian business processing industry is 'rising up the value chain', and while it may be that this aspiration is becoming a reality for certain kinds of back-office, non-customer-facing processes, as well as for IT and software services, the evidence suggests that the complexity of voice-based, call centre work remains limited. The following comment is typical of the tenor of contributions made in a debate on 'Moving Up the Value Chain: Non-core to Core' at the National Association of Software and Service Companies' (Nasscom's) annual ITES conference.

> When clients are coming in, they are looking at the situation and saying the experience that exists right now is pretty confined ... It probably takes a couple of weeks to train people for doing credit card applications, for doing collections work, or for doing basic customer service. This type of work will grow, but it is not really about the transition process [of core and complex work].
>
> (Vashistha, 12 June 2003)

Of course, given the fact that there is no shortage of this basic, transactional work in the call centres of the developed economies, it is not difficult to see how the Indian sector will continue to grow. However, this does not mean that transferring call centre work to India is seamless and non-problematic. In fact, the transnational character of the servicing relationship both intensifies and adds particular, culturally distinctive facets to generic difficulties common throughout the call centre industry.

Recruitment, selection and training

In the UK, North America and Australia, recruitment, selection and training have been identified as key sets of HR practice (for example, see Durbin this volume), particularly because tightly specified work regimes based on extensive monitoring and measurement cannot by themselves secure the desired quality service. Arguably, less emphasis is placed on technical proficiency and product knowledge than on positive attitudinal and communication skills (Callaghan and Thompson 2002). For example, conversation training is often concerned with mastering specific techniques (e.g. managing the conversation cycle) and controlling feeling towards customers (Thompson *et al.* 2004). Induction and training programmes are also geared heavily towards inculcating the norms and values of the company and customer service. Finally, given the central challenge of ensuring that employees perform emotional labour of consistently high quality, within this technologically driven, high volume, low-cost labour process, training aims to help agents adopt necessary coping mechanisms (Houlihan 2000, p. 231).

In the Indian context, while all these identified skills and competencies are required, the linguistic and cultural challenges involved in providing voice services add additional layers of complexity. For, although work-flows are highly standardised, it does not follow that the contradiction between the cost-reduction and customer-orientation logics can be resolved entirely in favour of the former. The *quality* of agent–customer interaction, albeit within the tightly prescribed parameters of standardised and mostly scripted calls, remains of great importance to the management of Indian service providers – whether these are third party outsourcers or 'captive' operations[3] – and to the offshoring company based abroad. No matter how standardised the call content, the inescapable requirement to communicate effectively, knowledgably and even profitably with customers in the West cannot be automatically guaranteed.

A key conclusion of our research is that, contrary to the received wisdom, it cannot be assumed that even 15 years of English education at school and university necessarily equips Indian workers with the ability to communicate with customers, for whom English is the mother tongue, to standards deemed acceptable by Indian managers and their overseas clients. Despite rigorous recruitment and selection procedures, as recommended by Nasscom (2002, p. 51), and extensive training programmes involving practices which Mirchandani (2003, p. 13) has defined as forms of linguistic and cultural imperialism, problems patently remain in achieving the desired quality of interaction between Indian agents and Western customers.

When recruiting, in addition to the familiar attitudinal characteristics sought (Callaghan and Thompson 2002), employers look for applicants, all of whom until recently were required to be university graduates, with the ability 'to grasp different accents and various cultural nuances or those who demonstrate "cultural fit"' (Scope Marketing 2001). Although Nasscom claimed (2003, p. 148) an offers-to-applicants ratio of 1:4 for the ITES industry, evidence from call centres specifically suggests a far higher level of failed applications of between 1:12 and 1:20. Following multi-stage screening, most failed to progress due to the thickness of their accents, which are reputedly stronger and less amenable to Anglicisation in certain regions of the country. Stringent selection occurs at an early stage because experience has informed Indian employers that customer service could be jeopardised should candidates, lacking the necessary standards of fluency, command and accent, progress to call-handling. Evidently, the selection of only those applicants with advanced English language capability is not always successful and has created problems both for the nascent Indian call centre industry and their overseas clients.

Admittedly, a high proportion – eight out of nine according to an IT/ITES survey (Nasscom 2003, p. 148) – of successful applicants accept the posts offered to them. The extent of take-up reflects both the large number of graduates, perhaps two million, who enter the Indian labour market annually, and the limited alternative job opportunities. Yet, in another challenge to popular perception, the labour market does not easily provide an endless flow of agents with the requisite language skills. In one survey, 28 per cent of call centres reported that they had experienced difficulty in recruiting agents (callcentres.net 2003, p. 35), with 38 per cent of outsourcing companies experiencing problems. Overwhelmingly, the biggest obstacle to recruitment, identified by 86 per cent of respondents, was the shortage of skilled agents in the marketplace. So great are the difficulties, exacerbated by high rates of attrition in a rapidly growing industry and the tendency for call centres to cluster, that certain state governments have intervened in attempts to overcome labour supply problems. For example, the Karnataka Government introduced a Skills Assessment Test in order to 'assess a person's communication skills, accent, clarity and grammar'.

Language, accent and cultural training are the aspects of Indian call centres that have received the greatest attention in the UK media. They are undeniably regarded by the industry as central priorities, for if agents are not able to communicate successfully with customers, calls might be abandoned or lost, misunderstandings might arise and, in extreme cases, customers might be lost to the organisation. In third party call centres, the insistence on high standards of spoken English is driven by the Western business clients, who are highly sensitive to customer resistance to offshoring, and the dissatisfaction that may follow perceptions of inferior service.

The transnational character of service provision ensures that there are two dimensions to accent training. Agents must be able to understand the national and regional accents of the English-speaking customers, while the latter must be able to comprehend the speech of the Indians who are servicing them. The most common practice is to 'neutralise' accents, to ensure, in the words of a Delhi manager, that agents 'do not speak with Indian inflections' (Interview, 24 January 2003). Considerable efforts are devoted to eliminating particular pronunciations which may be caused by regional accents. The authors observed a training session in which an expatriot Indian now resident in Canada was instructing inductees how to communicate with customers in North America. Following a role play exercise directed towards improving diction, she focused on students' pace of delivery, their tone and empathy and corrected their pronunciation. Repetition of apparently bizarre tongue-twisters aimed to rectify what Westerners might perceive to be imperfections in speech, in particular the pronunciation of consonants such as 'v' and 'r'. Yet, having neutralised accents, many call centres then encourage, or expect, their agents to adopt the accents of their customers. One manager stated that while the 'overall objective is to get CSRs (customer service representatives) to emulate received pronunciation' his company 'ideally want to get agents to speak with accents, although to expect this immediately would be fanciful' (Interview, 30 January 2003) and only through prolonged exposure to customers 'on the phone will agents get there'. Another Mumbai manager reported how the influence of television, films and music from the US led to agents acquiring slight American accents (Interview, 27 January 2003).

Modifying diction and natural rhythms of speech was combined with programmes designed to familiarise agents with the cultural background of customers' countries. Much of this cultural training is delivered by specialist outsourcers, such as Oceans Connect, a UK-based company which has recruited retired British schoolteachers to instruct Indians on the UK's political scene, its regional, economic and cultural differences, its principal cities and difficult place names (*Herald*, 6 August 2003). As Mirchandani (2003, p. 5) maintains, 'the very success of training programmes is signified by workers' ability to assimilate into ... (and identify with) American society while remaining within India'.

However, as we have argued elsewhere (Taylor and Bain 2005), despite these considerable efforts to modify speech and accent, and notwithstanding agents' formal command of the English language, overall linguistic capability *generally* is not sufficiently developed, or flexible, to guarantee that more than the most standardised call handling can be conducted from India. Of course, this is a provisional evaluation and the capacity for greater complexity will emerge but, for now, the limitations on moving to higher communication skills remain real. Similarly, it is questionable whether short-run acculturation programmes can bridge the distance between India and the West. Significantly, as our survey of UK and Scottish firms demonstrates, exploiting the educated, English-speaking human resources was cited as the second most important reason for offshoring, yet 'accent and language problems/cultural differences/mistakes through misunderstanding' were identified as the single most important problem experienced (Taylor and Bain 2003, pp. 56–58).

Labour process and task performance

The technological aspects of call centre work organisation in India are *essentially* the same as those experienced in the UK (see Taylor and Bain 2005 for a fuller discussion). Further, the mechanisms of control, in the form of the integrated technical, bureaucratic and normative controls that Callaghan and Thompson (2001) elaborated upon, are manifest in the call centres of the developed countries and are also found in the Indian workplace. That said, it is necessary to distinguish both subtle and significant differences in their application and effects. Generally, these distinctions are the product of the functional relationship of the offshored activity to its company and country of origin. As indicated in the previous discussion of the workflows which have been migrated, our evidence shows that the type of calls handled in India tends to be highly routinised, simplified in content, tightly scripted and of short duration. At one of the Delhi call centres investigated, tasks consisted of credit card debt recovery and collections, involving structured calls with cycle times of between three and four minutes. Managers at four of the five Mumbai call centres researched revealed that they undertook basic transactional services on behalf of US financial sector clients. Typical calls at one centre lasted between 30 and 180 seconds and this general simplicity in task performance is confirmed by the testimony of call handlers (Group discussion, 2 February 2003). Mirchandani (2005, p. 4) has confirmed how, for the most basic of workflows, foreign companies 'seamlessly outsource the exact scripts and procedures currently in place in their call centre'.

Thus, call centre work organisation most resembles an exaggerated version of the mass production model as delineated by Batt and Moynihan (2002). Consequently, agents' performance is extensively and intensively monitored and measured. Through contractual arrangements, similar in nature to those which apply in domestic outsourcing, service level agreements (SLAs) between supplier and client prescribe quantitative measures (including call volumes, abandonment rates, call-handling times and call waiting times). In turn, these quantitative measurements cascade throughout the call centre, ultimately becoming individual targets set for individual agents. As we have observed for the UK industry, quantitative targets are combined with the recording and monitoring of the content of agents' interaction with customers, which, in turn, produce additional qualitative targets (Bain *et al.* 2002; Bain and Taylor 2000; Taylor *et al.* 2002). Unlike much of the academic literature, which downplays the significance of targets, we contend that setting targets and systematically ensuring that agents meet them are central to management's labour utilisation strategy.

There are grounds for believing that performance measures are imposed with even greater rigour in India than in the UK. Survey data showed that concerns over managing from a distance were the second most frequently reported disadvantage by organisations who had already offshored (Taylor and Bain 2003, p. 57). Consequently, the need to overcome uncertainty and to exercise control over remote operations, ensuring the requisite quantity and quality of customer service, leads to extremely strict target adherence at all organisational levels. For example, Datta's (2004) fieldwork at three call centres reveals a claustrophobic intensity of controls, with call quality composed of up to 30 'parameters', many of which are subjective assessments of an 'associate's behaviour and manners'. All incoming calls were recorded and 12 calls per month, six each by the client and an 'internal analyst', were subjected to a quality of response analysis. The penalties for deviating from tightly prescribed norms could be punitive as one worker recalled:

> I must score 90 per cent on quality. Once, by mistake, I pressed a letter key instead of 'enter'. That changed the name of the customer. My score on quality became less than 90 per cent. I got into the 'bucket'. To come out of the 'bucket' I had to score more than 90 per cent for the next two consecutive months. Unfortunately, it took me 4 months to pull myself out of the 'bucket'. I already had been given by that time verbal and written warnings.
>
> (Datta 2004, p. 11)

Teamworking also appears to be utilised as a form of control, rather than genuinely interdependent, which involves the numerical aggregation of individuals undertaking a 'narrow range of routine and repetitive, largely scripted and technologically controlled, fragmented tasks' (Datta 2004, p. 12). Within and between teams, competitions for highest numbers of calls taken and lowest call-handling times (CHT) take place. However, competitive controls co-exist in teams alongside social and emotional control as team leaders counsel team members, not always successfully, in the demanding working environment.

Experiences of work

Evidence suggests that Indian call centre agents experience work as pressurised and frequently stressful. Nasscom has referred to 'high-volume induced burnout and the tedium of some of the tasks' (2002, p. 52). High turnover, now widely recognised to prevail throughout the industry, is perhaps the single most serious human resource problem. While the product of a rapidly expanding industry, it is also the consequence of working conditions that many find intolerable. For 'despite rather lucrative pay packets' (Nasscom 2002, p. 26) and limited alternative employment opportunities, many who exit a call centre leave the industry altogether. The employees in Datta's case studies (2004) remained with their employer, on average, for eight months. Datta (2004) found that discontent relating to work pressures was manifold amongst the 78 'associates' he interviewed.

Given what we know of the outcomes of call centre work where call throughput is prioritised and strict monitoring prevails (Deery *et al.* 2002; Taylor *et al.* 2003), it is unsurprising that, in India, where extreme standardisation occurs, employee exit and burnout appear commonplace. However, distinctive characteristics of the Indian industry exacerbate these generic problems. Call-handling for overseas customers takes place at night, or during evenings, on shifts that often last more than eight hours. In Datta's cases nine-hour shifts were the norm, but often were extended 'by another 3 to 4 hours for a "reward" of $6' (2004, pp. 7–8). 'Graveyard' shifts commencing in the early morning hours are also common. Managers identified the *combination* of night-time working and repetitive call-handling as stressful, taking its toll on agents' health and social and family life (Interviews, 27–30 January 2003). Long commuting distances compound these negative effects. Deb (2004, p. 20) reports his and fellow-workers' experiences of inhabiting 'a self-contained world of 13-hour days, taking into account the travelling time of up to two hours each way, with space for little else to penetrate one's existence'.

Agents confirm the prevalence of ill-health, with one typically commenting, that 'many cases have arisen where people have lost weight . . . usually because the cycle of eating and sleeping is disrupted. Symptoms include dark circles under the eyes . . . in most of the people' (Discussion, 2 February 2003). Thirty-three per cent of Datta's (2004, pp. 13–14) respondents reported difficulties in getting 'normal sleep', a problem intimately related to working long atypical shifts involving little task variation and autonomy. At a session at Nasscom's ITES-BPO Conference devoted to HR issues, participants expressed concerns regarding the effects of extended unsocial working on health and work–life balance. Women, who constitute half the workforce, are particularly affected by conflicts between working hours and the demands of task performance on the one hand, and domestic, family and social responsibilities and expectations on the other.

Although the built environment bears favourable comparison with UK facilities, the fact that buildings are sealed and that temperature, air conditioning and humidity are controlled by building management systems (BMS) means occupants are likely to experience problems. Lacking fresh air and opening windows, these artificially created microclimates have been linked by the World Health Organization to 'sick building syndrome' (HSE 1995). Symptoms could also be exacerbated by temperature and humidity extremes during the monsoon period, which BMS may be unable to counteract, particularly in circumstances of 24-hour building occupancy (Nasscom 2003, p. 63). Research (e.g. Baldry *et al.* 1997) has shown that widespread ill-health results from routinised labour processes taking place within unsupportive ambient and ergonomic environments.

The deleterious effects of task performance are aggravated by frustration, role confusion, a crisis of expectations and psychological tension experienced by Indian agents as they negotiate the contradictions between their culture, identity and aspirations, and the requirements of service provision for Western customers. The widespread adoption of anglicised pseudonyms, of having to conceal their Indian locations, and the obligation to speak in 'neutral' accents, or even emulate customer dialects, can contribute to a pressurised working experience. One Mumbai manager openly criticised these normative practices which, originating in GE's call centres, were spread throughout the expanding industry through a managerial diaspora:

> If agents are taking 150 calls a night from the east and west coasts [of the US], it puts even more pressure on them if they have to change from call to call to speak in the accent the customer uses. Customers can see through false accents.

Undoubtedly, the compulsion to conceal location – locational masking to use Mirchandani's term (2003, p. 16) – can increase tensions during encounters with customers. What motivates this deception is the assumption, shared by Indian providers and offshoring companies alike, that providing services from India produces customer backlash. Agents report how perceptions that Indian call centres give inferior service can lead to customer frustration, and even aggression, directed towards them as frontline staff (Datta 2004; Group discussion, 2 February 2003). As van den Broek (2004, p. 67) argues, 'one of the most invasive aspects of Indian call centre work is that many are expected to feign customer co-location by conversing about local weather conditions or cite aspects of recent sporting events'. When customers believe they are being lied to over location, anger can become overtly racist. In more subtle ways the performance of emotional labour is given added, culturally mediated twists which merely serve to intensify work pressure.

In addition, the gulf between the stimulating work and prestigious career promised by employers, and the mundane nature of incessant call-handling can cause disenchantment and disengagement from job and organisation. Articulating this conflict between career and status expectation on the one hand, and the reality of task performance on the other, some ex-graduates see work 'as demeaning and inappropriate given their qualifications' (Mirchandani 2003, p. 8). Undoubtedly, many agents leave to take up employment that may be less financially rewarding, but which appears more attractive and stimulating (Deb 2004, p. 18). Nevertheless, this should not lead to a one-sided conclusion to a phenomenon that is contradictory. As van den Broek (2004, pp. 68–69) suggests, the opening of call centre employment to layers of middle class women[4] is seen by many as creating opportunities and challenging traditional and restrictive cultural norms.

Workplace governance, employee relations and trade unions

There is some evidence that customary Indian hierarchical workplace cultures have been transposed to call centres (Outsourcing Insight 2001, p. 114). One associate in Datta's study revealed how, when call volumes are high, 'one has to raise a blue flag to attract the attention of the team leader for breaks and sometimes one cannot go to the loo' (2004, p. 7). Many workers evidently believe that they can be 'terminated', with little or no warning if, for example, they fall out with their team leader or manager or if they slip behind their targets. In the testimony of one worker:

> All it needs is for someone to be taken aside by a manager and told there is no future for him at the company and you won't see that person again. Everyone is so scared of being terminated that they will resign before they are pushed.
>
> (Cooke 2005, p. 10)

The reason for this is that if an agent is sacked, as opposed to resigning, he/she will not receive a 'release letter' making it virtually impossible to secure new employment.

As Nasscom's survey of the ITES-BPO sector indicated, top-down methods dominate companies' staff communication systems (Nasscom 2003, p. 149). Clearly, the employee involvement practices in place were task-based, geared to increasing efficiency, and were not intended to give employees a voice, let alone any real participation in decision-making. For example, while 88 per cent of companies claimed they had various policies for dealing with employee grievances, the most common practice was described as an 'open door' approach, but no formalised procedures were evident.

Despite strong trade union traditions in telecoms, banking (Kuruvilla *et al.* 2003) and insurance, there is no recognition for collective bargaining purposes, nor even embryonic union organisation in Indian call centres. There are several obvious reasons for this lacuna. First, call centres are a new development and established Indian unions have been slow to appreciate the significance of their emergence, making no real attempt to organise the workforces. Second, and relatedly, the Indian trade union officers that were interviewed by the authors saw call centres as barren ground, because of the widespread perception that staff were professionals who, coming from middle class backgrounds, would be unlikely to identify with collective workers' organisation. 'Giving the title of "Officer" and "Executive" to call centre workers ensures that they have a sense of status' (Interview, John 22 January 2003). The call centre workers interviewed concurred, seeing the inexperience, youth, professional identity and relatively high earnings of their fellow workers as disincentives to joining, or even looking favourably upon, unions. Third, there is the issue of employer opposition, whether this is the position of the third-party operator, or the policy of the 'captive'. Many of the 'captives' (GE, American Express) and the transnationals (IBM, EDS) who have more recently come to operate in the BPO sector have industrial relations' histories which demonstrate antipathy to trade unionism. Even at AXA at Bangalore which is perceived to be one of the better call centres and to have more favourably disposed employers, management 'was cool about the suggestion of an independent trade union' (Cooke 2005, p. 4).

It is surely significant that virtually all of the UK companies who have migrated call centre work to India recognise trade unions in the UK, but do not extend this to operations in India (Bain and Taylor 2002b; Taylor and Bain 2004). While there is no evidence to suggest that *explicit* union avoidance forms a central element in offshoring strategy, it is no doubt pragmatically beneficial, permitting labour flexibilities and contractual arrangements that would certainly be the subject of union negotiations 'at home' (Bain and Taylor 2004). Many UK companies, who have offshored, talk rhetorically of replicating identically their culture and working environment in India, yet union recognition, an essential element of employee relations, is not transferred. For this, and the other reasons outlined in this section, there is a democratic deficit in the workplace governance of Indian call centres.

The question of whether, and under what conditions, this might change is of considerable interest. It would appear that some of the more far-sighted employers in India are aware that the longer-term impact of call centre working conditions might be conducive to trade unionism. Sujit Baksai, the CEO of HCL Technologies-BPO Services Ltd, which purchased the unionised (CWU) British Telecom call centre in Belfast, believed that trade unionism might not be such a bad thing as 'it gives a certain discipline' (2003). However, in the Indian context the emergence of trade unions might not be so welcome.

> I strongly think that in my business call centres should have no more than 500 seats. Today, in Noida, I have three call centres of 600 seats each ... I think that having smaller units on a campus is important, because after all this business is very monotonous, very repetitive. If you look at the unions in the 1930s or the industrial revolution in Europe, any business where the job is monotonous – the automobile industry, the post and telegraph, banks – unless we create a variety of jobs, unions are bound to come.
>
> (Speech, 13 June 2003)

However, union growth is never automatic but depends, amongst other things, on the commitment of organisations to run sustained campaigns and upon the actions of union organisers (see Rainnie and Drummond this volume). Under the tutelage of UNI (Union Network International), significant progress has been made in several Indian cities in bringing together IT workers to discuss work-related problems in employees 'forum', with the intention of extending this form of organisation to embrace call centre workers (UNI Finance 2004). Examples of modest success have been reported. Cooke (2005, p. 6) recounts the story of a Delhi worker who organised a mini-rebellion at Wipro Spectramind against enforced extended working, lack of breaks and intimidatory management. Although eventually sacked, he established an email message group for disaffected BPO workers on Yahoo, which at December 2004, had 310 members. Important though these glimpses of resistance certainly are, trade unions are clearly some way from making inroads into the Indian BPO and call centre sectors.

Nevertheless, these instances of opposition suggest a general truth. It cannot be assumed that Indian call centre workers are simply more intensely exploited units of capital, but rather are active, social and cultural participants, who construct their own meanings of work (van den Broek 2004). They are not simply passive observers of the globalisation process and, even though open opposition to managerial authority might be limited, there is wider evidence of contestation in the workplace. At the most basic level, many agents demonstrate their lack of conformance by disengagement by exiting the workplace (see Taylor and Bain 2003, pp. 116–120 for a full discussion of attrition). In the workplace, Mirchandani (2003, p. 9) has documented how some workers clearly do know when supervisors are monitoring calls and adjust their behaviour accordingly, allowing them 'to maximise their [performance] scores'. Intriguingly, some manipulate procedures for resolving customer service problems in order to circumvent controls and maximise their bonuses. Future research will deepen our understanding of emerging forms of resistance but sufficient evidence already exists to demonstrate that not all Indian workers conform to the loyal and compliant stereotype propagated by Whitson (*Financial Times*, 6 August 2002) and Rice (*Sunday Herald*, 18 April 2004).

Conclusion

The character of work organisation in Indian call centres can be partly understood by reference to what we already know about call centres in the developed world. An extensive academic literature has provided us with valuable conceptual frameworks, and much case study and empirical evidence, which has deepened our knowledge. Nevertheless, there is a tendency to fail to embed workplace organisation and management control within wider political economies, leading both to over-optimistic characterisations of the work experience, and to strangely de-contextualised accounts. As Thompson (2003), has recently urged, this contextualisation is essential if we are to comprehend the principal drivers of change at the workplace level.

It is not the weightlessness of the new knowledge economy (Callinicos 2001; Huws 1999), or the arrival of a network society (Castells 2000), or the triumph of the consumer, which has created the situation where customers in the developed world are serviced by workers in the developing world. Admittedly, innovation and dramatic price reductions in information and telecommunication technologies have facilitated the growth of transnational servicing relationships. However, they are also the product of an era of deregulation, neo-liberalism and the growing assertiveness of multinational corporations. More importantly, though, just as the central capitalist imperatives of corporate cost-cutting, profit maximisation and shareholder pressure had proved influential in the widespread dissemination of the call centre as an organisational form, so too these have been the fundamental forces driving offshoring. If tightened labour markets in the 1990s were an initial stimuli, then the dot.com crash, the crisis of the new economy and the subsequent recession provided a sustained impetus to this attempt to find a 'spatial fix' to problems of cost and low profitability in customer servicing and back office work.

In this context, offshoring to India represents another stage, albeit a dramatic one, in the process of spatial relocation which, as we have seen, produced clusters of call centres in peripheral regions of the developed countries, notable for lower (principally labour) costs and the availability of skilled labour. The BPO industry and its call centre sector in India emerged, initially, as low cost replications of the most routinised of processes. Concretely, decisions at the level of the firm involved wholesale organisational restructuring, producing the segmentation of 'core' and 'non-core' processes, and the subsequent migration of some of the most standardised and lowest risk services. Ultimately, despite all the necessary caveats regarding quality, it was and is the cost-reduction logic that dominates, producing in India an extreme version of Batt and Moynihan's (2002) mass production model.

While the recognisable panoply of managerial controls is discernable, Indian national characteristics, particularly in the neo-colonial context, alter their content. As we argue elsewhere (Taylor and Bain 2005), the synchronisation of Indian agents' working times to customer servicing hours in the developed countries is both symbolic of the subordination of Indian workers to the interests of Western capital and, because it induces evening and nocturnal work patterns, significantly exacerbates the generic pressures associated with routine call-handling. Binding together Indian agents and Western customers in this transnational servicing (or selling) relationship generates additional contradictions and tension. Many Indian workers enter the Western-facing call centre with high expectations, believing much of the hype that they will be performing challenging and stimulating work on behalf of prestigious clients, or directly for well-known multinationals. While in a rapidly expanding industry there are those who are able to build a career to team leader and managerial levels, many conclude from the unwelcome realities of relentless call-handling, claustrophobic monitoring, strict conformance to the performance criteria demanded by overseas clients and long hours and travelling times that exit is a preferable option. Mirchandani (2005), particularly, has emphasised how locational masking, the compulsion (or encouragement) to adopt Western names and accents, can add culturally specific dimensions of psychological strain to the stresses frequently associated with emotional labour.

Furthermore, Indian call centre workers cannot readily find channels to articulate their concerns and channel their grievances. The Indian BPO industry remains union-free, and internal communication and employee involvement practices are simply mechanisms by which management can transmit its task-based commitment agendas. Nevertheless, it would be mistaken both theoretically and empirically to believe that Indian workers inevitably and willingly subordinate themselves to the imperatives of customer servicing as dictated by the interests of Western clients. Contestation may take familiar forms of opposition to oppressive working conditions and perception of injustice but, in subtle ways at micro-levels, it may involve conflict within the context of the cultural exchange, as Indian workers become the 'front line' representatives of UK and US companies.

The evidence contradicts the presumption that the transfer of call centre services to India is essentially non-problematic. The powerful cost-reduction logic[5] drives companies to offshore, but then they are confronted with problems deriving from the customer-oriented logic of call centre work. The need for Indian agents to communicate effectively with remote customers, whose mother tongue is English, however much that might be refracted by national or regional accent, presents enormous recruitment and training challenges. Even though the workflows are remarkably standardised and interactions heavily scripted, considerable efforts are devoted to ensuring that standards of spoken English and cultural assimilation are sufficiently advanced. It is not at all clear that these efforts are sufficiently successful in overcoming cultural and linguistic difference to encourage companies to offshore more than the most transactional processes. Of course, it must be conceded that this could change in the future and that more complex voice services might be migrated, for we must remember that we are only in the very first phases of what is a long-term process of the offshoring of services, whose outcomes are far from predetermined.

Notes

1 See Taylor and Bain (2003) for a full account of the methods and sources of data from this phase of the project.
2 For a full account of the development of India's ITES-BPO industry see Taylor and Bain (2003: 90ff) and (2004) for a summary.
3 One of the main distinctions that needs to be drawn in terms of the Indian call centre industry is that between 'captive' organisations – essentially operations catering for the internal requirements of MNCs such as GE Capital and American Express – and third party providers. Included in the latter category are dedicated BPO companies, operations created by Indian software companies, those created by traditional Indian business houses and, finally, and significantly, the global IT and BPO companies such as Accenture and EDS.
4 All the survey evidence suggests that equal numbers, or even a slim majority, of the Indian call centre workforce are women.
5 As we have demonstrated at length (Taylor and Bain 2003) there is evidence that labour costs are rising in India and that additional expenses include providing duplicate supplies of power and telecommunications. Nevertheless, so great is the cost differential between the UK and India that significant savings can be made by offshoring certain services.

References

Bain, P. and Taylor, P. (2000) 'Entrapped by the "Electronic Panopticon"? Worker Resistance in the Call Centre', *New Technology, Work and Employment*, 15(1): 2–18.

Bain, P. and Taylor, P. (2002a) 'Consolidation, "Cowboys" and the Developing Employment Relationship in British, Dutch and US Call Centres', in U. Holtgrewe, C. Kerst, and K. Shire (eds) *Re-Organising Service Work: Call Centres in Germany and Britain*, Ashgate: Aldershot.

Bain, P. and Taylor, P. (2002b) 'Ringing the Changes? Union Recognition and Organisation in Call Centres in the UK Finance Sector', *Industrial Relations Journal*, 33(3): 246–261.

Bain, P. and Taylor, P. (2004) 'No Passage to India? UK Unions, Globalisation and the Migration of Call Centre Jobs', paper presented to *Work, Employment and Society Conference*, Manchester, 1–3 September.

Bain, P., Watson, A., Mulvey, G. and Gall, G. (2002) 'Taylorism, Targets and the Pursuit of Quantity and Quality by Call Centre Management', *New Technology, Work and Employment*, 17(3): 154–169.

Baksai, S. (2003) Speech to *Nasscom ITES-BPO Strategic Strategy Conference*, Bangalore, 13 June.

Baldry, C., Bain, P. and Taylor, P. (1997) 'Sick and Tired? Working in the Modern Office', *Work, Employment and Society*, 11(3): 519–539.

Batt, R. and Moynihan, L. (2002) 'The Viability of Alternative Call Centre Production Models', *Human Resource Management Journal*, 12(4): 14–34.

Brenner (2002) *The Boom and the Bubble: the US in the World Economy*, London: Verso.

Bristow, G., Munday, M. and Griapos, P. (2000) 'Call Centre Growth and Location: Corporate Strategy and the Spatial Division of Labour', *Environment and Planning A*, 32: 519–538.

Cairncross, F. (1997) *The Death of Distance: How the Communication Revolution Will Change our Lives*, Boston: Harvard University Press.

Callaghan, G. and Thompson, P. (2001) 'Edwards Revisited: Technical Control and Call Centres', *Economic and Industrial Democracy*, 22(1): 13–37.

Callaghan, G. and Thompson, P. (2002) '"We recruit attitude": the selection and shaping of routine call centre labour', *Journal of Management Studies*, 39(2): 233–254.

callcentres.net (2003) *The 2003 India Call Centre Industry Benchmark Study*, Sydney: ACA Research.

Callinicos, A. (2001) *Against the Third Way*, London: Polity.

Castells, M. (2000) *The Rise of the Network Society*, Oxford: Blackwell.

Cooke, M. (2005) *Stretched to the Limit – CBPOP, BPOX and Call Centre Unionisation in India*, Report for Amicus, London: January 2005.

Datta, R. (2004) Worker and Work – a Case Study of an International Call Centre in India, paper to *22nd International Labour Process Conference*, Amsterdam, 5–7 April 2004.

Deb, S. (2004) 'Call me', *Guardian Weekend*, London: 3 April, 14–23.

Deery, S., Iverson, R. and Walsh, J. (2002) 'Work Relationships in Telephone Call Centres: Understanding Emotional Exhaustion and Employee Withdrawal', *Journal of Management Studies*, 39(4): 471–496.

Dicken, P. (2003) *Global Shift*, London: Sage.

Financial Times (2002) 6 August.

Fransman, M. (2002) *Telecoms in the Internet Age*, Oxford: Oxford University Press.

Frenkel, S., Korczynski, M., Shire, K. and Tam, M. (1999) *On the Front Line: Organisation of Work in the Information Economy*, Ithaca: Cornell University Press.

Harman, C. (1996) 'Globalisation: a Critique of a New Orthodoxy', *International Socialism*, 2: 73.

Harvey, D. (1982) *The Limits to Capital*, Oxford: Oxford University Press.

Harvey, D. (2000) *Spaces of Hope*, Edinburgh: Edinburgh University Press.

Held, D., McGrew, A., Goldblatt, D. and Perraton, J. (1999) *Global Transformations – Politics, Economics and Culture*, London: Polity Press.

Henwood, D. (2003) *After the New Economy*, New York: The New Press.

Herald (2003) 6 August.

Hirst, P. and Thompson, G. (1996) *Globalisation in Question*, Oxford: Oxford University Press.

Houlihan, M. (2000) 'Eyes Wide Shut? Querying the Depth of Call Centre Learning', *Journal of European Industrial Training*, 24(2–4): 228–240.

Houlihan, M. (2002) 'Tensions and Variations in Call Centre Management Strategies', *Human Resource Management Journal*, 12(4): 67–85.

HSE (1995) *How to Deal with SBS: Sick Building Syndrome – Guidance for Employers, Building Owners and Building Managers*, London: Health and Safety Executive Books.

Hutchinson, S., Purcell, J. and Kinnie, N. (2000) 'Evolving High Commitment Management and the Experience of the RAC Call Centre', *Human Resource Management Journal*, 10.1: 63–78.

Huws, U. (1999) 'Material World and the Myth of the "Weightless Economy"', in L. Panitch and C. Leys (eds) *Socialist Register 1999*, London: Merlin Press.

Kohler, G. (2003) 'Foreign Direct Investment and Its Employment Opportunities in Perspective: Meeting the Great Expectations of Developing Countries?', in W. Cooke (ed.) *Multinational Companies and Global Human Resource Strategies*, Westport: Quorum.

Korczynski, M. (2002) *Human Resource Management and Service Work*, Basingstoke: Palgrave.

Kuruvilla, S., Frenkel, S. and Peetz, D. (2003) 'MNCs as Diffusers of Best Practice in HRM/IR in Developing Countries', in W. Cooke (ed.) *Multinational Companies and Global Human Resource Strategies*, Westport: Quorum.

Marshall, J. N. and Richardson, R. (1996) 'The impact of "telemediated services" on corporate structures: the example of "branchless" retail banking in Britain', *Environment and Planning A*, 28: 1843–1858.

Mirchandani, K. (2003) 'Making Americans: Transnational Call Centre Work in India', Unpublished paper.

Mirchandani, K. (2005) 'Webs of Resistance in Transnational Call Centres: Strategic Agents, Service Providers and Customers', in R. Thomas, A. Mills and J. Helms-Mills (eds) *Gender, Organisation and the Micro-politics of Resistance*, London: Routledge.

Nasscom (2002) *IT Enabled Services:Background and Reference Resource*, New Delhi: Nasscom.

Nasscom (2003) *Strategic Review: The IT Industry in India,* New Delhi: Nasscom.

Ohmae, K. (ed.) (1995) *The Evolving Global Economy: Making Sense of the New World Order,* Boston: Harvard Business Review Press.

Outsourcing Insight (2001) *Call for India,* London: Outsourcing Insight.

Richardson, R. and Belt, V. (2001) 'Saved by the Bell? Call Centres and Economic Development in Less Favoured Regions', *Economic and Industrial Democracy,* 22(1): 67–98.

Ruigrok, W. and van Tulder, R. (1995) *The Logic of International Restructuring,* London: Routledge.

Scope Marketing (2001) *Call Centre Industry,* Chennai: Scope Marketing.

Sunday Herald (2004) 18 April.

Taylor, P. and Bain, P. (1997) *Call Centres in Scotland: A Report for Scottish Enterprise,* Glasgow: Scottish Enterprise.

Taylor, P. and Bain, P. (1999) '"An Assembly Line in the Head": Work and Employee Relations in the Call Centre', *Industrial Relations Journal,* 30(2): 101–117.

Taylor, P. and Bain, P. (2001a) 'Trade Unions, Workers' Rights and the Frontier of Control in UK Call Centres', *Economic and Industrial Democracy,* 22(1): 39–66.

Taylor, P. and Bain, P. (2001b) *Call Centres in Scotland in 2000,* Glasgow: Rowan Tree Press.

Taylor, P. and Bain, P. (2003) *Call Centres in Scotland and Outsourced Competition from India,* Stirling: Scotland.

Taylor, P. and Bain, P. (2004) 'Call Centre Offshoring to India: the Revenge of History?', *Labour and Industry,* 14(3): 15–38.

Taylor, P. and Bain, P. (2005) 'India Calling to the Far Away Towns: the Call Centre Labour Process and Globalisation', *Work, Employment and Society,* 19(2): 261–282.

Taylor, P., Hyman, J., Mulvey, G. and Bain, P. (2002) 'Work Organisation, Control and the Experience of Work in Call Centres', *Work, Employment and Society,* 16(1): 133–150.

Taylor, P., Baldry, C., Bain, P. and Ellis, V. (2003) '"A Unique Working Environment": Health, Sickness and Absence Management in UK Call Centres', *Work, Employment and Society,* 17(3): 435–458.

Thompson, P. (2003) 'Disconnected Capitalism: or Why Employers can't Keep their Side of the Bargain', *Work, Employment and Society,* 17(2): 359–378.

Thompson, P., Callaghan, G. and van den Broek, D. (2004) 'Keeping up Appearances: Recruitment, Skills and Normative Control in Call Centres', in S. Deery and N. Kinnie, (eds) *Call Centres and Human Resource Management,* Basingstoke: Palgrave, pp. 129–152.

UNI Finance (2004) Available at: www.union-network.org/unifinance.nsf> (accessed 26 May 2004).

Vanaik, A. (2004) 'Rendezvous at Mumbai', *New Left Review,* 26: 53–65.

van den Broek, D. (2004) 'Globalising Call Centre Capital: Gender, Culture and Work Identity', *Labour and Industry,* 14(3): 59–75.

Vashistha, A. (2003) Speech to *Nasscom ITES-BPO Strategic Strategy Conference,* Bangalore, 12 June.

Venkatraman, R. (2003) Speech to *Nasscom ITES-BPO Strategic Strategy Conference,* Bangalore, 12 June.

4 German call centres between service orientation and efficiency

'The polyphony of telephony'

Claudia Weinkopf

Introduction

In recent years call centres have been one of the fastest growing areas of employment in Germany. Call centres tend to be characterised by a pronounced division of labour. This reflects their basic philosophy of removing customer contacts from the case handling process, concentrating them into specific organisational units and dealing with them solely by telephone to the greatest possible extent. Tasks are often prescribed as highly standardised, frequently monotonous and with limited time allowed for completion. It is against this background that call centre work is frequently described as the 'neo-Taylorism' of service activities (D'Alessio and Oberbeck 2002). However, previous research on call centres in Germany (briefly summarised by Holtgrewe 2003) and in several other countries (e.g. Batt 2000; Kinnie *et al.* 2000; Frenkel *et al.* 1999; Wray-Bliss 2001) has illustrated that it is inappropriate to lump all call centres together. Call centres are far from being easily characterised by badly paid, monotonous and simple service work (see for example Koskina's chapter on Greece in this volume).

Moreover, such simplification should be avoided for a number of reasons: First, the type and complexity of services rendered by call centre employees varies considerably. They range from simple telephone directory information to demanding customer consultancy services and skilled clerical operations, and even medical as well as legal specialist advice. Second, the frequent use of twofold motivation by companies for setting up or using call centres is another factor; that is, improving their customer services on the one hand and reducing the cost of those services on the other. This obviously results in a variety of HRM strategies with different outcomes in terms of working conditions. Third, call centre work can be regarded as interactive service work, being characterised by the decisive influence through which the behaviour and competency of personnel is exerted on the quality of services rendered (Rieder and Matuschek 2003). As a result, it is impossible for management to influence individual interactions directly. Rather, management have to design framework conditions so that employees will behave in the best interests of 'their' company. A wealth of call centre literature indicates that 'commitment and control' are the two poles between which HRM strategies at call centres move.

These issues raise questions concerning how German call centres master the balancing act between cost efficiency and service quality in practice and what implications this has on HRM strategies, job quality and employees' job satisfaction. In this chapter, these issues will be discussed against the background of a review of the findings of available studies and surveys in Germany – most of them carried out since the end of the 1990s and frequently based on company case studies. This also applies to the FREQUENZ study which was funded by the German Federal Ministry of Education and Research and was conducted by the Institut Arbeit und Technik and B+S Management Consultants in co-operation with 18 call centres and four companies in the retail sector, in the period from February 2000 to January 2002. Originally, the main focus of the study had been the strategies developed by the companies (in call centres and in retail trade) in order to cope with the fluctuations in customers' demand. However, a literature review (Bittner *et al.* 2000) demonstrated that there was limited information on the HRM strategies of German call centres at that time, so it was decided to study those issues more broadly. The sample for the subsequent study comprised a wide range of centres that differed considerably in size, organisational structures and services provided. In all companies, case studies focusing on several features of work organisation and HRM strategies were carried out by conducting interviews with managers, supervisors, staff level planners and works councils or employees. The FREQUENZ study also included a number of workshops with representatives of the call centres in which several issues such as employee satisfaction were discussed. In this context, it was decided to carry out two written surveys among agents and team leaders – both of these surveys being implemented and analysed by B+S Management Consultants.[1] The employee survey included 650 employees in 14 call centres, however, not all of these call centres were involved in the case studies (John and Schmitz 2002).

After a short overview of the structure and evolution of the German call centre market, various aspects of job quality, employees' satisfaction, HRM strategies and industrial relations will be discussed. Finally, the prospects and perspectives of work within call centres in Germany will be considered. Findings will suggest that the prevailing pessimism regarding the working conditions at call centres seems inappropriate.

The German call centre market

The German call centre market emerged strongly in the 1990s in response to both technological developments and commercial pressures. Around 80

Table 4.1 Basic data on call centres in Germany

Year	No. of call centres	No. of employees	No. of seats
1998	1,600	150,500	79,200
1999	2,300	187,300	98,600
2000	2,750	224,800	108,300
2001	3,350	261,800	137,800
2002	3,750	280,000	150,000
2003	4,300	320,000	162,000
2004	4,900	330,000	170,000

Source: Deutscher Direktmarketing Verband 2004.

per cent of all call centres in Germany were established since 1992. As in other countries, it is almost impossible to trace their quantitative development owing to the absence of official data on the industry. Data on the number and structure of call centres is principally fraught with insecurities and thus inconsistencies, because there is not a uniform and precise definition of what is understood by the term 'call centre' (Arnold and Ptaszek 2003). It is hardly surprising, therefore, that multi-source quantitative data on call centre distribution and development is not consistent with the size of the industry. Table 4.1 presents recent estimates for the development of the call centre industry in Germany (including the number of employees and seats) over the past seven years.

The sectors in which German call centres operate are similar to those in other countries. German call centres tend to cluster around towns (Arnold and Ptaszek 2003, p. 41), as this is often where a parent company may be located (although this is not a requirement for technical reasons). Industrial conurbations offer the advantage that a bigger workforce with required qualifications is available (EIRR 2000, p. 14). However, several German call centres – frequently providing rather simple services – are also located in structurally weak rural areas with high unemployment rates. These centres tend to be partly supported by public aid programmes (Arzbächer *et al.* 2002, p. 34).

By the end of the 1990s, the share of in-house call centres in Germany was estimated at between 60 and 70 per cent (Bittner *et al.* 2002, p. 66), whereas more recent surveys suggest that the majority are now being represented by independent or outsourced service providers (SoCa 2004; Holtgrewe 2005). It is questionable, however, as to how far this relates to a substantial growth of this type of call centre or is due to the fact that it is much more difficult to identify in-house centres, which may lead to an under-estimation of their prevalence. An increasing share of independent or outsourced service providers in Germany might be partly due to the fact that such an 'externalisation' of services offers the opportunity to escape from institutional contexts, especially from collective agreements (Shire *et al.* 2002, p. 6; Kerst and Holtgrewe 2003, p. 85). In-house companies are also buying external call centre services, in addition to their own activities, in order to cope with peaks of call volume and to cover certain times in the evenings and at weekends to process simple matters. The latter can also be a strategy of call centre managers to preserve better working conditions for their own employees (Bittner *et al.* 2002, p. 92).

Job quality and job satisfaction

Job quality is determined by numerous factors, where statistical significance and the direction of effect may partly exhibit case-related differences (see for example Meisenheimer 1998, p. 23f.). The European Commission (2001, p. 65) gives the following definition:

> Job quality is a relative concept regarding a job–worker-relationship, which takes into account both objective characteristics related to the job and the match between worker characteristics, on the one hand, and job requirements, on the other. It also involves subjective evaluation of these characteristics by the respective worker on the basis of his or her characteristics, experience, and expectations. In the absence of a single composite indicator of job quality, an empirical analysis of job quality necessarily has to be based on data on both objective job and worker characteristics and subjective evaluations of the job–worker match.

The following analysis also includes both objective working conditions at German call centres and their employees' individual assessments, focusing on some central aspects of job quality in call centres (pay, working time, workload, work contents, opportunities for further training and advancement) and demonstrating the variety of findings in these fields.

Overall satisfaction

The overall job satisfaction of call centre agents who participated in the FREQUENZ employee survey was at 73.4 per cent, whereby 14.6 per cent stated that they were very satisfied, and 58.8 per cent that they were satisfied. The dissatisfaction rate, by contrast, was only 8.3 per cent (of which 1.4 per cent indicated that they were very dissatisfied). The categories selected for the survey were not directly comparable with the classification chosen for a study of the European Commission (2001, p. 66) (high – medium – low overall satisfaction), however, they pointed in a similar direction. In Europe, the rate of employees with low overall job satisfaction was indicated at 7.7 per cent, with the share of those indicating high job satisfaction amounting to 50.9 per cent.[2]

A differentiated analysis of the results yielded by the FREQUENZ survey of call centre agents showed that a considerable spread or dispersion of satisfaction values in the various companies lingered behind the – on average – relatively high overall satisfaction of respondents. The lowest values were 45.5 per cent, whereas in the company with the highest satisfaction rate, as many as 84.6 per cent of the respondents stated that they were all in all satisfied or even very satisfied (John and Schmitz 2002, p. 34).

Pay

Call centre work is generally considered to be an area with relatively low remuneration. This does not only apply to Germany, but also to many other European countries (EIRR 2000, p. 15). In the call centres covered by the FREQUENZ study, the average entry-level salary of call centre agents in full-time employment amounted to €1,693 gross per month, which corresponded to an hourly wage of approximately €10. However, the spread between the companies was extremely wide: earnings in the call centre with the best pay were at an hourly gross rate of €19 per hour, and were thus three times as high as earnings at the call centre with the lowest pay of €6 per hour.

A more recent survey of 154 German call centres has led to similar results (Holtgrewe 2005) where the mean hourly wage of all call centres was €11.30 (which corresponds to around three-quarters of the median hourly wage in Germany). The highest mean remuneration of €13.19 was paid by in-house call centres and the lowest of €10.47 was paid by independent service companies. The differences become even more obvious by differentiating with regard to the question of whether a collective agreement is applied or not. In call centres not covered by collective agreements the mean hourly wage was only €10.54, whereas in call centres with a collective agreement pay was a mean hourly wage of €13.72 (industry-wide agreement) or even €14.27 (company agreement).

In line with the extreme spread of earnings at the call centres the wage satisfaction rates in the companies varied considerably, ranging between 6.1 per cent and 72.3 per cent (John and Schmitz 2002, p. 34). On average, only 8.0 per cent of the respondents stated that they were very satisfied with their income, and 34.5 per cent indicated that they were satisfied. The share of those who claimed that they were dissatisfied with their wage level amounted to 29.3 per cent (of whom 7.1 per cent were very dissatisfied).

Working time and employment arrangements

Call centres generally have a large share of part-time workers. In the 18 FREQUENZ companies, the average part-time share was 43 per cent, which is exactly the rate present in the most recent industry survey (Holtgrewe 2005). As can be expected, there were considerable gender-specific employment differences: the female part-time rate amounted to 56 per cent and was higher than the male part-time share (27 per cent). In many cases, the part-time rates differed considerably between the various call centres (Figure 4.1). The sample encompassed both call centres that employed full-time workers, either exclusively or predominantly, and call centres operating almost exclusively on part-time work – apart from the managerial level. What should be emphasised in this context is that the high part-time rate of above 80 per cent was consistently found only in call centres that were providing comparatively simple services, such as tele-phone directory information and order intakes. Conversely, the more complex and qualified the services provided, the higher the rate of full-time employment (Bittner *et al.* 2002, p. 69f).

Furthermore, call centres generally have extended operating times, compared with other industries, and working-time organisation tends to be flexible, because one important goal is to adjust personnel deployment as perfectly as possible to the (expected) call volume. This is one reason for the relatively high part-time employment share. Nevertheless, the so-called 'mini-jobs'[3] with very short individual working times and monthly earnings of up to €400, which are particularly widespread in several other service sectors in Germany such as retail, commercial cleaning and hotels, seem to be unimportant in call centres. This may be related to the fact that call centres frequently invest considerably in their employees' training and this would not pay off in the case of short-time deployments of personnel (Bittner *et al.* 2002, p. 247; Kerst and Holtgrewe 2003, p. 93).

The question as to how far call centre employees experience flexible working times as a strain and burden naturally depends to a large extent on whether they can influence the spread and timing of their shifts, and on the notification period within which actual working times are announced.

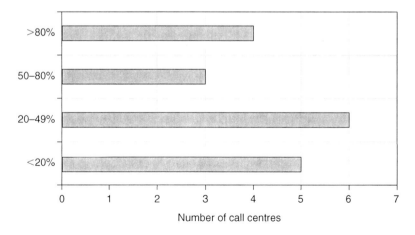

Figure 4.1 Ratio of part-time workers in the FREQUENZ call centres (*n* = 18).

This period was at least two weeks in the FREQUENZ call centres, and it was normally possible to swap shifts with colleagues (Bittner *et al.* 2002, p. 90f). Working-time satisfaction among those agents who took part in the employee survey was above average, compared with other aspects tested for satisfaction (31.1 per cent very satisfied and 45.2 per cent satisfied) (John and Schmitz 2002, p. 61).

With regard to employment practices, it is notable that German call centres do not use the entire range of flexible and contingent working arrangements. According to the results of the most recent industry survey (Holtgrewe 2005), the intake of freelancers and temporary agency workers tended to be rather low (averaging at around 5 per cent); 51 per cent of all call centres used permanent contracts only, whereas 3 per cent relied on fixed-term contracts exclusively. In our sample, the rate of fixed-term contracts among the employees was 10.4 per cent, which is quite close to the overall average in Germany (Bittner *et al.* 2002, p. 66).

Workloads and work contents

In line with common assumptions and findings, the employee survey results demonstrate that any room for manoeuvre for call centre agents is rather limited. Only 22.1 per cent of the respondents agreed fully or mostly with the statement that they had a say in the allocation of work. The share of those who stated that they have the possibility to co-design their direct work environment was only slightly higher, amounting to 27.3 per cent (John and Schmitz 2002, p. 57).

Leaving aside periods where staff undertake post-processing work, or other telephone-related work tasks (the general features of work at call centres) the survey results indicate as Taylor and Bain (1999, p. 110) argue that '... even in the most quality driven call centre it is difficult to escape the conclusion that the labour process is intrinsically demanding, repetitive and, frequently, stressful'.

Training and career development opportunities

It is often assumed that the education level of call-centre workers is relatively low (EIRR 2000, p. 14). For Germany at least, this cannot be confirmed. From the call centres involved in the FREQUENZ study, 91 per cent of the employees had completed vocational training within the German dual system of apprenticeship comprising three years of study (Bittner *et al.* 2002, p. 72). Other studies have indicated similar results (e.g. Baumeister 2001; Böse and Flieger 1999; Holtgrewe 2005). At the same time, many call centres indicate that applicants' social-communicative skills hold more weight in personnel recruitment than vocational knowledge. Call centre organisations attempt to account for both technical and social skills by offering various levels of induction for new employees in addition to on-the-job training and coaching (Bittner *et al.* 2002, p. 106ff). In this context, Holtgrewe and Kerst (2002) draw attention to the fact that German call centres offer a considerable amount of organisational reflection and investment in relation to quality management for subordinate and medium-qualified labour. This is also corroborated by the results of the employee survey: 62.6 per cent of agents stated that training and coaching measures were implemented in their companies on a regular basis, and 65.9 per cent agreed with the statement that the 'breaking in' of new employees was prepared and implemented with care. Once again, however, there were considerable differences between the various companies: the rates of agreement varied from 4.3 to 92.8 per cent. Furthermore, the share of those who enunciated a big interest in further training and in the acceptance of more responsibility was extremely high at 80.6 per cent (John and Schmitz 2002, p. 29).

The Achilles heel of call centre work is, without doubt, the rather limited career opportunities. This circumstance is often due to the 'flat hierarchies' of call centres, which often operate with three hierarchical levels (agents, team leaders/supervisors and management), and the low level of work differentiation (Bittner *et al.* 2002, p. 119ff). Only 27.7 per cent of the employees stated that they had good career prospects in their company (John and Schmitz 2002, p. 56). Since the rise to leadership positions is open to only a small number of agents, and since the possibilities to change over to other corporate areas are very limited, some call centres have created an additional management level featuring deputy team leaders. Other companies place more emphasis on the so-called horizontal competence careers. These are characterised by the assumption of additional tasks in the areas of job familiarisation and training, or quality control (Bittner *et al.* 2002, p. 114ff). Such approaches can alleviate the problem of weaker career opportunities, but they cannot remedy and eliminate the problem altogether.

Design options

The brief analysis of several features of call centre work in Germany has illustrated that job quality and employee satisfaction differs considerably. There are variable outcomes with respect to working conditions, remuneration, working times, qualifications and career prospects. However, relatively few differences emerge with respect to workloads and work content. Despite providing different services, the core work activity consists of the handling of telephone enquiries in all call centres. Obviously, the work-organisational decision to bundle the handling of customer enquiries within a call centre and leave follow-up and post-processing activities either to the back office or to other corporate departments almost inevitably leads to a certain one-sidedness in the agents' work experience.

To avoid this, a more comprehensive work design seems to be required.

The German debate suggests that there are various possibilities to this effect, which revolve around a reduction of telephone-centred time.[4] Instead, employees should be given other tasks corresponding to the same, or an even higher level of qualification, such as the handling and processing of written enquiries by email, fax, letter or text messaging, which, in the process of developing call centres into 'customer care centres', are gaining importance in many companies. Other options include involving call centre agents more strongly in planning, steering and controlling tasks, in order to alleviate the division of labour towards the back office and in order to introduce partly autonomous teamwork with a much larger scope of decision-making for call centre agents (Mola and Zimmermann 2001, p. 23).

The combination of telephone enquiry handling with other non-telephone-centred tasks might also bring further advantages for call centre companies. Notably, the often highly sophisticated personnel planning procedures required to predict fluctuations in the volume of work and the deployment of the appropriate number of workers over the course of a day or a week, as precisely as possible, would be facilitated (Bittner *et al.* 2002, p. 85).

Such a strategy was being implemented in one of the FREQUENZ companies, namely in the 'virtual' call centre of a health insurance company. There, the case-handling clerks of selected offices were each responsible for dealing with telephone enquiries as part of their job. There was no open-plan office with many telephone work stations, instead the employees undertake their call-centre work at their normal place of work (Bittner *et al.* 2002, p. 140). It is questionable, however, whether this will be a trend-setting approach for other companies that emphasise high service quality or whether it will remain an exception.

Different types of HRM practices

Although a wide range of different features and procedures is to be found within German call centres, there are certain shared characteristics which can be condensed into three basic types of HRM strategies.[5] The factor that exerts the strongest influence on HR practices is the *type of service*. Our main distinction is between the two poles being 'complex' on the one hand, and 'mass' on the other hand. To characterise the third type in the middle of the following chart, the description 'demanding and flexible' has been used. As Table 4.2 shows, the three types appear in different prominence also with regard to a range of other criteria.[6]

In using the above typology, the structural characteristics of call centres have intentionally been left aside. However, if these are included in a second step, it becomes clear that at least some of their features correlate strongly with the various types of HR practices. Obviously, the *company's background* plays a major part in the type of HRM pathway pursued:

Table 4.2 Types of personnel management in call centres

	Type 1 'complex'	Type 2 'demanding and flexible'	Type 3 'mass'
Type of service	Complex and consultancy-intensive	Moderately difficult to complex	Mostly simple
Collective bargaining commitment	Yes	No	Partly yes, partly no
Share of telephone-centred working time	Rather low	Medium	Rather high
Service times	Rather short	Extensive	Average
Remuneration	Above average	Average, partly performance-related wage elements	Below average
Full-time ratio	High	Medium	Low
Qualification requirements	Mostly technical	Technical and social-communicative	Mostly social-communicative
Working times	Fixed working times or flexitime	Fixed working times or shift systems	Shift systems
Strategic corporate orientation	Focus on service quality	Service quality and cost efficiency	Focus on cost efficiency

- The type *'complex'* exclusively comprises call centres that have developed from 'traditional' industries, such as the insurance business and the chemical industry. The call centres represented in this group are all in-house call centres. Each of them was founded as part of an existing company organisation and aims at speeding up and improving the company's reaction to changed requirements in customer communication. The recruitment of agents is effected mainly from the existing workforce. Moreover, the clear focus of all of these call centres is on inbound telephony.

- The type *'mass'* encompasses all those call centres providing rather simple and standardised telephone services, which have been offered for decades: telephone directory enquiries, mail orders or telephone operator services. However, it is not necessarily 'traditional' companies that take over these tasks. New companies may just as well have been called into being in the wake of outsourcing processes or in terms of a new corporate establishment. Consequently, both in-house call centres and external service providers are located in this category. Furthermore, the category contains call centres with a focus on either inbound or outbound activities.

- The medium type *'demanding and flexible'* covers call centres which provide services to new industry segments or which render novel services, respectively – for example direct banking, EDP support or telemarketing. It is the characteristic of these call centres that they are either external service providers or in-house call centres that form part of an outsourced company. None of these companies is protected by collective bargaining agreements. Usually, they do not have a works council in their firm, or – if they have one – it has been set up recently. The conditions of work are often worse than in companies of the 'complex' type.

It is particularly the 'hard criteria' – such as employment structures, work organisation and wage levels – which contribute to the distinction of HRM types, whereas hardly any differences could be detected in matters of personnel development and leadership. This finding suggests that there is definitely room for manoeuvre when it comes to the design of HRM strategies in call centres.

Industrial relations

Call centres comprise a relatively new work field which is still under construction, and thus, teems with 'work in progress'. Such circumstances make specific demands on trade unions. Call centres are not an autonomous industry, but a form of organisation gaining ground in very different industries and contexts. In the case of an industry-related organisation of trade unions, such as in Germany, the question concerning the scope of competence and responsibility comes into play (Arzbächer *et al.* 2002, p. 27; Kutzner and Kock 2003, p. 179). A further difficulty lies in the fact that call centres are frequently set up in order to provide services at a lower cost. Outsourcing, or making use of independent service providers, offers the opportunity to undermine collectively agreed standards of the parent company (for example in financial services – see Hild 2003, p. 83; Körs *et al.* 2003, p. 159).

At the end of the 1990s, trade unions estimated the proportion of call centres with collective agreements in Germany to be approximately 50 to 60 per cent (Meier 1999). In the sample of the most recent survey, the share of call centres covered by collective agreements was even smaller, at around 26 per cent (Holtgrewe 2005), which might be partially due to the relatively low share of in-house call centres involved. So far, the large majority of outsourced call centres and independent service providers are not being covered by collective agreements in Germany. According to the same survey, works councils were established in 45 per cent of the companies, and thus more frequently represented, than collective agreements. In this case, the difference between in-house and external or outsourced service providers was also obvious. While there was a works council in approximately 84 per cent of the in-house call centres, this applied to only 21 per cent of the independent service companies. Outsourced subsidiaries were ranked in between with a share of 50 per cent.

As union representation (in terms of collective agreements and the presence of works councils) is substantially higher among German in-house call centres, it can be concluded that works councils should continue trying to avoid the outsourcing of call centres. However, there are also several signs suggesting that trade unions and works councils are not without opportunities in the newly established call centres, either (see Rainnie and Drummond this volume). One of the driving forces seems to be the widespread feeling of unfair treatment among employees and the desire for more transparency (Kutzner and Kock 2003, p. 175f).

The findings also suggest that the organisation of employee representation tends to be easier than the setting up of collective agreements. So far, German trade unions have only been successful in setting up company-related collective agreements and the chances for agreements based on industry-wide standards – such as those concluded in Sweden in February 2001 for the call centres and telemarketing companies (see EIROnline 2001)[7] – tend to be low. This is due to several reasons – ranging from the lack of employer associations willing to negotiate about standards, to the low rate of union membership among call centre employees, and the divided responsibilities among the industry-related trade unions in Germany.

Conclusions

The intention of this chapter was to provide insight into what is going on in German call centres, with particular attention being paid to job quality, employee satisfaction, HRM strategies and industrial relations. It has been established that there is a broad variety of features and practices that suggest that call centres are not, in every case, the 'sweat shops' or 'dark satanic mills' of the twenty-first century where workers are recklessly exploited (see Fernie and Metcalf 1998; Knights and McCabe 1998).

Nevertheless, in the context of the rather comprehensive German labour market regulations and institutions, call centres seem to be characterised by a marginal attachment to this framework. Collective agreements are quite rare in call centres, works councils are not very widespread and, to date, there is no specific vocational training for call centres within the German dual system of apprenticeship. Moreover, to some extent, the growth of the German call centre market might even be enforced by offering a way to escape from traditional regulations in order to provide services at lower cost (D'Alessio and Oberbeck 2002, p. 97).

It would, however, be a mistake to regard this area as beyond the pale of trade union organisation and to assume that the institutional framework has hardly any impact. Outside of exempt part-time employment, call centres do not make use of flexible and contingent working arrangements to a large extent. This is a notable distinction from several other service occupations in Germany in which large proportions of staff are employed in 'mini-jobs' which are frequently related to low pay. The broad majority of call centre employees have completed vocational training and particular training schemes for call centre work have been developed and implemented. The extent of monitoring and control seems to be less comprehensive than in other countries such as the UK (Taylor and Bain, 1999). This may be related to the relatively wide range of legal options of works councils to exert influence on the shaping of substantial working conditions such as performance-related monitoring (Shire *et al.* 2002, p. 8). The share of German call centres with a solely cost-cutting orientation and rigidly controlled and standardised work organisation is, therefore, assumed to be lower than that of several other countries (Holtgrewe 2003, p. 51).

All in all, the odds for an improvement of working conditions at German call centres are not bad, the more so as the companies themselves are still seeking solutions to many of the issues raised in the research (Kerst and Holtgrewe 2003, p. 91) and a trend towards higher-quality services can already be detected. However, as several publications have indicated (see Matuschek and Kleemann 2002), this tends to be a necessary, but not sufficient, precondition for the emerging relevance of high road HRM practices. Against this background, future research will need to investigate, in more detail, under which circumstances better working conditions in call centres can be realised.

Notes

1 In the team leader survey, 133 team leaders of 47 call centres were involved (Schmitz 2001). One consequence of this initiative was the establishment of a team leader network and the organisation of annual conferences since 2001.

2 The empirical results of the European Commission are based on data from the European Community Household Panel (ECHP, 1994–1996) with more than 120,000 observations per year. See European Commission 2001: 65.

3 Since April 2003 the regulation of the marginal part-time jobs in Germany has been modified – leading to a substantial increase of their number up to around 7 millions. Restrictions on the weekly working time have been eliminated and the monthly wage threshold has been raised substantially up to €400. For employees, these jobs are not subject to the usual social security contributions and income tax, whereas employers now have to pay a flat rate of 25 per cent of the monthly wage, which is a bit higher than the regular rate of social security contributions (about 21 per cent). However, they often compensate for these extra costs by paying lower wages or by excluding the employees from regular benefits.

4 Furthermore, this offers an appropriate opportunity of implementing the German Decree On Work With Display Terminals: the decree stipulates that companies have to organise work with display terminals in a way that workloads are reduced, either by interspersing other activities or by regular paid breaks (Mola and Zimmermann 2001, p. 22).

5 This is especially true of the two 'pole-side' types comprising five call centres each which are also intrinsically homogeneous. In the medium type by contrast, which encompasses the remaining eight call centres, some features differ in prominence. However, they are distributed very differently so that a further subdivision of this type did not seem useful.

6 Our typification bears a strong resemblance to two further typifications developed at the University of Duisburg (Gundtoft and Holtgrewe 2000, p. 184f; Kerst and Holtgrewe 2003). Wallace *et al.* (2000, p. 180) also distinguish in their study between 'transaction centres' (high efficiency, low service level), 'sales centres' (balance between service and efficiency) and 'solutions centers': 'Here the focus is completely on the customer in terms of access, personalised attention and quality of advice.'

7 This framework collective agreement contains provisions on working-time flexibility, the promotion of stable employment and compensations on overtime.

References

Arnold, K. and Ptaszek, M. (2003) 'Die deutsche Call-Center-Landschaft: Regionale Disparitäten und Arbeitsmarktstrukturen', in F. Kleemann and I. Matuschek (eds) *Immer Anschluss unter dieser Nummer. Rationalisierte Dienstleistung und subjektivierte Arbeit in Call Centern,* Berlin: edition sigma, pp. 31–47.

Arzbächer, S., Holtgrewe, U. and Kerst, C. (2002) 'Call Centres: Constructing Flexibility', in U. Holtgrewe, C. Kerst and K. Shire (eds) *Re-Organising Service Work. Call Centres in Germany and Britain,* Aldershot: Ashgate, pp. 19–41.

Batt, R. (2000) 'Strategic Segmentation in Frontline-services: Matching Customers, Employees and Human Resource Systems', *International Journal of Human Resource Management,* 3: 540–561.

Baumeister, H. (2001) *Call Center in Bremen. Strukturen, Qualifikationsan-*

forderungen und Entwicklungstendenzen, Herausgegeben von der Arbeitnehmerkammer Bremen.

Bittner, S., Schietinger, M., Schroth, J. and Weinkopf, C. (2000) *Call Center – Entwicklungsstand und Perspektiven. Eine Literaturanalyse*, Projektbericht des Instituts Arbeit und Technik 2000–01, Gelsenkirchen.

Bittner, S., Schietinger, M. and Weinkopf, C. (2002) *Zwischen Kosteneffizienz und Servicequalität – Personalmanagement in Call Centern und im Handel*, Schriftenreihe des Instituts Arbeit und Technik, Band 22, München/Mering: Hampp.

Böse, B. and Flieger, E. (1999) *Call Center – Mittelpunkt der Kundenkommunikation. Planungsschritte und Entscheidungshilfen für das erfolgreiche Zusammenwirken von Mensch, Organisation und Technik*, Braunschweig/Wiesbaden: Vieweg.

D'Alessio, N. and Oberbeck, H. (2002) 'Call Centres as Organisational Crystallisation of New Labour Relations, Working Conditions and a New Service Culture?'. in U. Holtgrewe, C. Kerst and K. Shire (eds) *Re-Organising Service Work. Call Centres in Germany and Britain*, Aldershot: Ashgate, pp. 86–101.

Deutscher Direktmarketing Verband (2004) *Wirtschaftsfaktor Call Center*, www.ddv.de/downloads/WirtschaftsfaktorCallCenter.pdf (accessed 25 February 2005).

EIROnline (2001) 'First Collective Agreement Signed for Call-centres and Telemarketing', www.eiro.eurofound.ie/print/2001/02/inbrief/SE0102183N.html (accessed 27 May 2005).

European Commission (2001) *Employment in Europe 2001. Recent Trends and Prospects*, Luxemburg.

European Industrial Relations Review (EIRR) (2000) 'Call Centres in Europe: Part One', *European Industrial Relations Review*, 320/September: 13–20.

Fernie, S. and Metcalf, D. (1998) 'Not Hanging on the Telephone: Payment Systems in the New Sweatshops', CEP-discussion paper No. 390, London: Centre for Economic Performance.

Frenkel, S. J., Korczynski, M., Shire, K. and Tam, M. (1999) *On the Front Line: Organisation of Work in the Information Society*, Ithaca: ILR Press.

Gundtoft, L. and Holtgrewe, U. (2000) 'Call-Center – Rationalisierung im Dilemma', in: H.-G. Brose (ed.) *Die Reorganisation der Arbeitsgesellschaft*, Frankfurt: Campus, pp. 173–203.

Hild, P. (2003) 'Call Center in markt- und ressourcenbasierter Strategieperspektive', in F. Kleemann and I. Matuschek (eds) *Immer Anschluss unter dieser Nummer. Rationalisierte Dienstleistung und subjektivierte Arbeit in Call Centern*, Berlin: edition sigma, pp. 65–84.

Holtgrewe, U. (2003) 'Call-Center-Forschung: Ergebnisse und Theorien', in F. Kleemann and I. Matuschek (eds) *Immer Anschluss unter dieser Nummer. Rationalisierte Dienstleistung und subjektivierte Arbeit in Call Centern*, Berlin: edition sigma, pp. 49–61.

Holtgrewe, U. (2005) *Call Centres in Germany – Preliminary Findings from the Global Call Centre Project – Germany*, Report for the Russell Sage Foundation (April 2005), draft, Duisburg.

Holtgrewe, U. and Kerst, C. (2002) 'Call Center: Die Institutionalisierung von Flexibilität', *Industrielle Beziehungen*, 9: 186–208.

John, H. and Schmitz, E. (2002) *Mitarbeitermotivation im Call Center. Ergebnisse der Mitarbeiterbefragung 2001*, Ergebnisse Forschungsprojekt FREQUENZ –

Personalmanagement Call Center und Handel, Herausgegeben von B+S Integrative Unternehmensberatung, Band 2, Bonn.

Kerst, C. and Holtgrewe, U. (2003) 'Interne oder externe Flexibilität? Call Center als kundenorientierte Organisationen', in F. Kleemann and I. Matuschek (eds) *Immer Anschluss unter dieser Nummer. Rationalisierte Dienstleistung und subjektivierte Arbeit in Call Centern*, Berlin: edition sigma, pp. 85–107.

Kinnie, N., Hutchinson, S. and Purcell, J. (2000) '"Fun and Surveillance": the Paradox of High Commitment Management in Call Centres', *International Journal of Human Resource Management*, 5: 967–985.

Knights, D. and McCabe, D. (1998) '"What Happens when the Phone Goes Wild?": Staff, Stress and Spaces for Escape in a BPR Telephone Banking Work Regime', *Journal of Management Studies*, 2: 163–194.

Körs, A., Von Lüde, R. and Nerlich M. R. (2003) 'Zwischen Computer Telephony und Voice Recognition – Zur Zukunft des Human Aspect in der Teledienstleistungsarbeit', in F. Kleemann and I. Matuschek (eds) *Immer Anschluss unter dieser Nummer. Rationalisierte Dienstleistung und subjektivierte Arbeit in Call Centern*, Berlin: edition sigma, pp. 141–161.

Kutzner, E. and Kock, K. (2003) 'Zur Strukturierung von Arbeitsbeziehungen in Call Centern', in F. Kleemann and I. Matuschek (eds) *Immer Anschluss unter dieser Nummer. Rationalisierte Dienstleistung und subjektivierte Arbeit in Call Centern*, Berlin: edition sigma, pp. 163–181.

Matuschek, I. and Kleemann, F. (2002) 'Arbeitsmotivation und langfristige Perspektiven in High-Quality Call Centern', in E. Kutzner and K. Kock (eds) *Dienstleistung am Draht – Ergebnisse und Perspektiven der Call Center Forschung. Sfs Dortmund, Beiträge aus der Forschung*, 127: 95–101.

Meier, C. (1999) 'Strategien zur Verbesserung der Arbeitsbedingungen in Call Centern aus gewerkschaftlicher Sicht'. Vortrag anlässlich des Pressegesprächs Arbeit im Call Center: Arbeitsplatz der Zukunft mit Schattenseiten der Redaktion, Computer-Fachwissen für Betriebs- und Personalräte' am 15 Juni 1999 in Frankfurt. Unveröffentlichtes Manuskript.

Meisenheimer, J. R. II (1998) 'The Service Industry in the "good" versus "bad" jobs debate', *Monthly Labour Review*, 2: 22–47.

Mola, E. and Zimmermann, E. (2001) 'Arbeitsorganisation im Call Center – Chancen einer menschenzentrierten Arbeitsgestaltung', in: *Arbeitsorganisation im Call Center: Teamarbeit mit qualifizierten Beschäftigten*, Veranstaltungsdokumentation der Hans-Böckler-Stiftung, des Kooperationsbüros Multimedia und Arbeitswelt und der TBS beim DGB NRW e.V.

Rieder, K. and Matuschek, I. (2003) 'Kritische Situationen in Dienstleistungstransaktionen', in F. Kleemann and I. Matuschek (eds) *Immer Anschluss unter dieser Nummer. Rationalisierte Dienstleistung und subjektivierte Arbeit in Call Centern*, Berlin: edition sigma, pp. 205–222.

Schmitz, E. (2001) *Teamleiter im Call Center, Ergebnisse der Teamleiterbefragung 2001/Teamleitertag 2001*, Ergebnisse Forschungsprojekt FREQUENZ – Personalmanagement Call Center und Handel, Herausgegeben von B+S Integrative Unternehmensberatung, Band 1, Bonn.

Shire, K., Holtgrewe, U. and Kerst, C. (2002) 'Re-Organising Customer Service Work: An Introduction', in U. Holtgrewe, C. Kerst and K. Shire (eds) *Re-Organising Service Work. Call Centres in Germany and Britain*, Aldershot: Ashgate, pp. 1–16.

SoCa (2004) *Die soCa-Beschäftigtenbefragung – ausgewählte Daten*, Soziale Gestaltung der Arbeit in Call Centern, Bremen.

Taylor, P. and Bain, P. (1999) '"An Assembly Line in the Head': Work and Employee Relations in the Call Centre', *Industrial Relations Journal*, 2: 101–117.

Wallace, C. M., Eagleson, G. and Waldersee, R. (2000) 'The Sacrificial HR Strategy in Call Centres', *International Journal of Service Industry Management*, 2: 174–184.

Wray-Bliss, E. (2001) 'Representing Customer Services: Telephones and Texts', in A. Sturdy, I. Grugulis, and H. Willmott (eds) *Customer Service. Empowerment and Entrapment*, Houndmills: Palgrave, pp. 38–59.

5 A national survey of Korean call centres

Byoung-Hoon Lee and Hye-Young Kang

Introduction

The business model of call centres, equipped with information and communication technologies (ICT), has been diffused across borders as a global trend. This is identified in the noticeable employment growth of the call centre industry within advanced countries. Moreover, the rapid growth of the call centre workforce has been viewed as a significant part of transforming employment and industrial structures in North America and Western Europe over the past decade (Holtgrewe *et al.* 2002). As a consequence, there has been a proliferation of research literature on call centres within those regions.

Similarly, the call centre industry in Korea, which originated from customer service offices handling customers' complaint calls in the late 1980s, has been booming during recent years (particularly since 1998), owing to the country's intense development of ICT infrastructure. According to a rough estimation of the Callcenter Information Research Centre (2004), the number of call centres in Korea grew from 2,000 in 2002 to 2,500 in 2004, while the employment size of this sector increased from 250,000 to 330,000[1] in the same period. There are good prospects for the country's call centre sector showing a 'compressed growth' of over ten per cent during the latter half of the decade from 2000, leading the market volume of call centre business to double from US$5.2 billion to US$10 billion between 2002 and 2007. This may be attributed to call centres being introduced as a key means of customer relations management (CRM) in many industrial sectors, including a variety of private and public services and manufacturing (Lee *et al.* 2004). In particular, telecommunications service, finance and electronic appliance manufacturing are the leading sectors actively utilising customer service contact of call centres.

In contrast to the accumulation of research findings on call centres in advanced economies, there is little research covering this burgeoning sector in developing countries. As a result, the authors aim to fill a research vacuum concerning the call centre sector of the developing economies whereby this chapter delineates both the operational and employment relations characteristics of Korea's call centres.

The survey

This chapter draws upon a call centre survey conducted between June and August, 2004. The survey is the first national survey concerning call centres' operation and employment relations in Korea. It is also conducted as part of the global benchmarking survey of human resource practices and performance in call centres, which is co-managed by the Global Call Center Industry Project team comprising 15 countries, including the US, the UK, Germany and Japan. The global call centre benchmarking (GCCB) survey, which was devised as a consequence of the development of call centre research in such advanced countries as the US and the UK, comprises 10 sub-sections:

- overview of call centre operation
- employee composition
- payment systems and performance appraisal
- training and development
- recruitment, staffing and employment relations
- performance monitoring
- job design
- operational performance
- call centre technologies
- institutional context.

During February–March 2004, the authors conducted a pilot test of the standard GCCB survey applying to five call centres and modified some of the survey questionnaires in order to reflect the unique features of Korea's employment relations practices.

Since no official data exist relating to the call centre population in Korea, all the available call centre information was collated by referring to telemarketing associations, Internet job openings for call centre representatives and call centre management forums. Approximately 400 companies were identified that were operating call centres in this stage of data collection. Two hundred and fifty companies operating call centre(s) with over ten customer service representatives (CSRs) were selected as survey candidates, since above this size the organisations would require a systematic employment relations programme.

Senior managers in charge of call centre operations were contacted by phone and approval to conduct the survey was obtained from 127 companies. Forty-one companies allowed a face-to-face survey with call centre managers, while in 86 cases they were returned by email or mail. By excluding four cases for not providing sufficient data, 123 call centres were included in the analysis. This sampling is approximate but provides a good starting point towards an understanding of the Korean call centre industry.

Operational characteristics and employment structure of call centres in the survey

Table 5.1 presents a general overview of call centres in the survey. The average employment size of call centres is 196.1 employees. Sixty-five per cent of call centres have fewer than 100 employees, whereas only approximately ten per cent have over 500, including four 'mega call centres' with over 1,000 employees. Those mega centres are affiliated to telecommunications service (three) and electronic manufacturing (one) sectors. The call centre sector in Korea has a short history of 'late development', in that the majority (58.4 per cent) of companies have established call centres since 2000, while only 15.7 per cent of call centres have operated over the previous ten years. The survey includes call centres in a wide range of industries. The industrial sector with the largest representation of call centres is financial services (44 centres or 35.8 per cent of the total), including distribution banking, insurance and credit cards, which was followed by distribution/hotel/restaurants (24 centres or 19.5 per cent) and manufacturing (21 centres or 17.1 per cent). The surveyed companies were operating 2.1 call centres on average, and 62.6 per cent of them have only one call centre. Fifteen per cent of the companies were running more than

Table 5.1 Overview of call centres in the survey

	Categories	*Frequency (%)*
Employment size (n = 123)	Less than 30 employees	42 (34.1)
	30–99 employees	38 (30.9)
	100–499 employees	31 (25.2)
	500 employees and above	12 (9.8)
Establishment year (n = 89)	1995 and before	14 (15.7)
	Between 1996 and 1999	23 (25.8)
	2000 and after	52 (58.4)
Industrial sector (n = 123)	Manufacturing	21 (17.1)
	Distribution, hotel, restaurants	24 (19.5)
	Telecommunications and transport	12 (9.8)
	Finance and insurance	44 (35.8)
	Private service, media, entertainment	13 (10.6)
	Public service and government	9 (8.3)
Number of call centres (n = 123)	One	77 (62.6)
	Two or three	27 (22.0)
	Four and above	19 (15.4)
Type of call centre (n = 123)	In-house centre	89 (72.4)
	Subcontractor	34 (27.6)

three call centres, and four firms (two each in telecommunications service and financial services) have more than 10 call centres. Seventy-four per cent of call centres were in-house operations, while 27.6 per cent were outsourced to subcontractors specialised in call centre operation services. The sectors of telecommunications service (75.0 per cent), public service (55.6 per cent), and manufacturing (42.9 per cent) have a higher use of outside subcontractors than other industries.

From Table 5.2, over 90 per cent of call centres in the survey locate their operational focus on in-bound call service, while 12 call centres (9.9% of the total) provide more outbound call services than inbound call services. The outbound service-based call centres mainly comprised financial (six) and manufacturing (three) sectors. Almost two-thirds (66.1 per cent) of call centres assume customer service, handling billing or service inquiries, as the most significant function of their business. The number of installed work stations at call centres ranged from four to 4,900 and the average was 213.9 units. The number of active work stations was between four and 4,800, and the average was 190.9 units. Fifty-one per cent of call centres have a daily operating schedule for customer service representatives to provide call services for between ten and 19 hours per day, while 21 call centres (17.2 per cent) offer a 24 hour service. The average working hours of CSRs was 42.8 hours per week, and their working schedule ranged from 31.5 hours to 57 hours per week.

As illustrated in Table 5.2, 75.6 per cent of call centres cover the entire country in service provision, while 16.3 per cent focus on particular provinces. Fifty-one per cent of call centres target the general mass retail market, while only two call centres focus exclusively on the corporate business market. Twenty-six and 13.8 per cent of call centres target multiple customer segments – the first for mass and VIP customers, and the latter for all types of customers, including mass, VIP and corporate business. The largest group of respondents (42.5 per cent) gave the highest priority to the fostering of customer loyalty as their business strategy. Another 26.7 per cent focus on service differentiation by getting their staff dedicated to a particular customer segment or a particular service. During the last two years (2002–2003), 29.5 per cent (29 call centres) increased business revenue, while only 8.2 per cent (eight centres) experienced a decrease in operational revenue; 44.9 per cent of call centres reported that their total revenue was unchanged during this period.

Table 5.3 shows that the employment size of call centres varies substantially by industry. The telecommunications and transport industries have the largest employment size with 723.8 persons per centre, while the sectors of private service, media and entertainment are the smallest, averaging only 27.3 persons per centre. The employment composition at call centres on average is 6.4 managers, 11.2 supervisors and 178.5 CSRs per centre. The relative ratios of CSRs per manager and supervisor are respectively 27.9 and 15.9, but these numbers vary by industry. Telecommunica-

Table 5.2 Operational characteristics of call centres

	Categories	Frequency (%)
Relative share of inbound call service (n = 121)	75% and above	69 (57.0)
	Between 50 and 74%	40 (33.1)
	Below 50%	12 (9.9)
Key contents of call service (n = 112)	Customer service (for inquiries, billing, etc)	74 (66.1)
	Operator services (e.g. directory assistance)	8 (7.1)
	IT help desk	7 (6.3)
	Telemarketing	6 (5.4)
	Collections (on past due accounts)	6 (5.4)
	Others	11 (9.8)
Number of work stations (or seats)	Average number of installed seats (n = 119)	213.9
	Average number of operating seats (n = 121)	190.9
Daily operating hours of call centre (n = 122)	Below 10 hours	38 (31.1)
	Between 10 and 19 hours	62 (50.9)
	20 hours and above	22 (18.0)
Scope of call service market (n = 123)	Local	8 (6.5)
	Regional	20 (16.3)
	National	93 (75.6)
	International	2 (1.6)
Target customers (n = 123)	General (mass market) customers only	63 (51.2)
	General and VIP customers both	32 (26.0)
	Corporate customers only	2 (1.6)
	General and corporate customers both	9 (7.3)
	General, VIP and corporate customers	17 (13.8)
Business strategy of call centres (n = 120)	Price leadership	7 (5.8)
	Service differentiation	32 (26.7)
	Customer loyalty	51 (42.5)
	One-stop service bundling	16 (13.2)
	Brand identification	6 (5.0)
	Others	8 (6.7)

tions and transport (47.1) have the largest ratio of department managers to CSRs. Finance and insurance (22.1) are the sectors having the highest relative ratio of CSRs per supervisor, followed by manufacturing (18.2).

The composition of CSRs differs between in-house and subcontracted call centres, in terms of employment type. Non-regular (temporary and

Table 5.3 Employee composition of call centres on average

	Manager	Supervisor	CSR	Sub-total
Manufacturing (21)	6.1	6.4	116.4	128.9
Distribution, hotel and restaurants (24)	4.2	9.2	140.4	153.8
Telecommunications and transport (12)	14.0	49.8	659.9	723.8
Finance and insurance (44)	7.6	7.3	161.4	176.3
Private service, media, entertainment (13)	1.1	2.2	24.1	27.3
Public service and government (9)	4.0	7.0	70.3	81.3
Total (123)	6.4	11.2	178.5	196.1

Note
Number of responding cases within the parentheses. CSR denotes customer service representative.

Table 5.4 Employment type of customer service representatives

	In-house call centres	Subcontractor call centres
Regular employees (%)	17.1 (9.2)	92.8 (60.9)
Non-regular employees (%)	169.3 (90.8)	59.8 (39.2)
Temporary workers	74.5 (44.0)	13.6 (22.7)
Part-timers	6.4 (3.8)	9.9 (16.6)
Contracted/dispatched workers	84.8 (50.1)	34.4 (57.5)
Self-employed/independent contractor	3.6 (2.1)	1.9 (3.2)
Total (%)	186.5 (100)	152.5 (100)

fixed term) employment of CSRs comprises 90.8 per cent for in-house call centres and 39.2 per cent at subcontractor call centres, as illustrated in Table 5.4. In light of the fact that subcontractor call centres are outsourced from the customer contact function of client companies, the employment of Korean call centres is 'externalised' to an extreme extent. When combining non-regular employees of in-house centres and all employees of subcontractor centres, 93 per cent of total CSRs are externalised in various forms of non-regular employment. Among those non-regular work patterns, indirectly employed labour, consisting of temporary agency help and contracted workers, comprises over 50 per cent at both in-house and subcontracted centres.

Table 5.5 summarises the requisite personal attributes of call centre employees as per the results of the survey questionnaire. Call centre employees in Korea are highly educated with 67.2 per cent of CSRs and 85.6 per cent of managers possessing a two-year college education at the minimum.[2] The average age of CSRs and managers was respectively 27.7

Table 5.5 Personal attributes of call centre employees

	Customer service representative	Manager
Education		
High school graduates (%)	32.8	14.4
Graduates of 2-year college (%)	38.5	20.1
4 year college and above (%)	28.7	65.5
Average age (years)	27.7	35.1
Average service years	Regular employees: 3.2	6.1
	Non-regular employees: 1.6	
Relative share of female employees (%)	91.0	55.1

and 35.1 years. The average tenure of CSR was 3.2 years for regular employees and 1.6 years for non-regular employees. This partly reflects the short history of the call centre sector in Korea. As in Western countries (Belt 2002; Mulholland 2002; Buchanan and Kock-Schulte 2000; Taylor and Bain 1999), the clear majority (91 per cent) of CSRs (and 55.1 per cent of managers) are female. Moreover, 30.3 per cent of CSRs have prior experience of call centre work, which implies that there exists a call-centre labour market exemplified by CSRs' job mobility within this sector. The average daily absenteeism rate at call centres was 4.6 per cent and yearly turn-over for CSRs in 2003 was 18.4 per cent.

Human resource management of call centres

As illustrated in Table 5.6, Korean call centre managers value CSRs' organisational loyalty and competency rather than their short-term performance. This is despite their commitment to labour cost savings, the heavy use of non-regular labour and outsourced subcontractors. This also reflects the recognition of call centre management that CSRs' interactions with customers are very significant for the success of corporate business, as pointed out by Houlihan (2002) and Frenkel *et al.* (1999).

Approximately 90 per cent of call centres have HR and training departments at the workplace level (see Table 5.6). The number of HR and training department staff at call centres was respectively 2.51 and 3.45 on average. At the call centres with those HRM-related functions, each member of staff of the HR and training department covers 62.2 and 45.6 CSRs, respectively.

In 2003, call centres hired 58.1 CSRs on average. The average selection rate was 35.7 per cent, about one CSR for every 2.8 applicants. Almost half of call centres (48.2 per cent) in the survey plan to increase the number of CSRs in future, which was indicative of the potential of the call centre industry to create jobs. Also, 30.8 per cent of call centres reported that they experienced some difficulty in recruiting new CSRs.

Table 5.6 Human resource management of call centres

HRM staff	Average number of HR department staff	2.51 ($n=99$)
	Average number of training department staff	3.45 ($n=99$)
Characteristics of HRM	Focus on (1) labour cost saving → (7) organisational loyalty	5.16 ($n=100$)
	Staffing by (1) external recruitment → (7) internal development	5.34 ($n=98$)
	Relying on (1) non-regular labour → (7) regular employment	4.13 ($n=98$)
	Emphasis on (1) short-term performance → (7) long-run competency-building	4.82 ($n=97$)

Note
Characteristics of HRM are measured by 7-point scale.

In recruiting new CSRs, the most common method was through Internet job posting (60.4 per cent), followed by private employment services (18.8 per cent) and personal referral of existing call centre employees (14.9 per cent). The fact that public employment service channels were hardly used indicates that public job-matching infrastructure for the industry is not well-developed and that management takes responsibility for recruiting new CSRs without any policy assistance.

Over 50 per cent of call centre managers regarded job applicants' attitude and team spirit as the most important element of selection criteria (see Table 5.7). Traditional human relations-oriented criteria were considered to be the primary factor in selecting new CSRs, which was consistent with call centre managers emphasis on HRM policy to foster team spirit and employee loyalty. On average, call centres used 2.54 testing tools to examine CSR applicants. The majority of call centres made use of interview tests (99.2 per cent) and curriculum vitae review (96.6 per cent) for selecting new CSRs, with approximately 40 per cent also adopting a simulation test of job ability. In 2003, call centres spent approximately US$110 on recruitment costs for each new CSR.

As summarised in Table 5.8, the length of entry-level training for new CSRs was 20 days on average. While 42.5 per cent of call centres offered initial training of ten days or less, over 30 per cent provided training for over 20 days. Transportation and public power utilities, in particular, made the largest investment in initial training (34.8 days and 33.5 days). At the other end of the spectrum, private service, media and entertainment provided only 11.8 days of entry-level training. This wide variation of initial training by industry can be explained by the degree of call service complexity. Overall, entry-level training cost call centres about US$2,617 per employee in 2003.

Call centre managers estimated that it took about 11.7 weeks for CSRs to become proficient in their jobs. While the finance sector, including

Table 5.7 Selection of customer service representatives (CSR)

	Categories	Frequency (%)
Recruiting channel for hiring new CSRs (n=101)	Personal referral of call centre employees	15 (14.9)
	Job opening notice by media	2 (2.0)
	Job opening notice by Internet	61 (60.4)
	School employment service office	2 (2.0)
	Public employment service agency	2 (2.0)
	Private employment service agency	19 (18.8)
Key consideration of CSR selection (n=76)	Job skills and voice	22 (29.0)
	Call centre job experience	13 (17.1)
	Job attitude and team spirit	41 (53.9)
Testing methods for CSR selection (n=119)	Curriculum vitae review	115 (96.6)
	Interview test	118 (99.2)
	Psychometric and aptitude test	22 (18.5)
	Simulation test of job ability	47 (39.5)
	Average number of testing methods	2.54

Table 5.8 Training of customer service representatives

Average length of entry-level training (n=120)	20.0 days
Average period for CSR's job proficiency (n=120)	11.7 weeks
Average length of on-going training for experienced CSRs (n=116)	18.1 days
Average frequency of training programmes by training type	
Updates on product or service information (n=116)	4.03
Customer interaction skills (n=119)	3.71
Interpersonal or team-building skills (n=116)	2.86
Stress management (n=116)	2.49

Note
Frequency of training programmes is measured by a five-point scale (1=none, 2=a few, 3=a moderate amount, 4=a lot, 5=a very great deal).

banking and insurance, required the longest work period (14.2 weeks) for job proficiency, the public sector needed only 8.2 weeks. Ongoing training was a significant part of HRM practice at the call centres. On average, call centres offered 18.1 days of ongoing training per year for experienced CSRs. Ongoing training was mainly task-oriented, aimed towards providing updated information of products or services and enhancing the CSRs' communication skills to interact with customers. However, the provision of training to improve employees' social skills and stress management was only moderate (see Table 5.8).

Call centres report that CSRs on average earned US$15,846 in 2003. However, the pay level of CSRs varied considerably by employment type and industry, ranging from US$6,000 to 40,000. The average annual pay of regular CSR employees was US$18,654, while that of non-regular CSRs was US$14,804. The annual pay of CSRs serving the public sector was the highest, averaging US$ 21,283, while for the manufacturing sector it was only US$13,842. The average pay of CSRs in the other industrial sectors was between US$15,406 and 16,936. The pay level of managers was US$35,437, more than twice as much as that of the average salary for CSRs. Just over 46 per cent of call centres in the survey implemented performance-based pay incentives, through individual commission or group bonuses. The fact that the average labour cost occupies 68.8 per cent of the total operating expenses at call centres indicates that this business sector is labour-intensive.

Call centres in Korea have a 'segmented career ladder', evidenced by the limited possibilities for CSRs to get promoted to higher positions (Lee *et al.* 2005). Fifty-four per cent of surveyed call centres reported that CSRs can be promoted to supervisory positions, while 39.6 per cent allowed supervisors to advance to managerial positions. That is, only 34 per cent of call centres guarantee CSRs a managerial career. This implies that non-regular, female representatives at two-thirds of call centres were excluded from the opportunity of job promotion. Most call centres conducted a formal performance appraisal on a scheduled basis with 83.4 per cent of CSRs on average being assessed by regular performance evaluation, and 76.1 per cent of call centres applied formal appraisal to all CSRs on the payroll.

Most call centres operated a union-free workplace with only 8 per cent of in-house call centres represented by labour unions. The low unionisation of the call centre sector can be explained by the predominance of a female and non-regular workforce, organisational independence from existing unions and management's strong union avoidance strategies. Non-union employee representative bodies (17.4 per cent) and labour–management councils (23.4 per cent) had a presence at a minority of call centre workplaces. Almost 80 per cent of call centres had a grievance procedure to resolve individual employees' concerns. Clearly, Korean call centre managers prefer to deal with employees individually, as opposed to dealing with collective organisations.

Work organisation of call centres

Electronic monitoring of CSRs at work is a common feature of call centre operations (Batt *et al.* 2004). With regard to the Korean call centre survey results it was found that 63 per cent of an average CSRs' work activity was monitored in a real-time manner (see Table 5.9). At more than two-thirds of call centres, supervisors listened to CSRs' calls daily or several times per week. Forty-four per cent of call centres delivered the results of work monitoring to CSRs as supervisor feedback and coaching on-call service technique several times per week; 38.5 per cent undertake the process

Table 5.9 Work monitoring of customer service representatives (%)

Average percentage of work activity of core employees monitored in a real-time manner (*n* = 112)		63.0
How often supervisors listen to CSR's calls (*n* = 116)	Daily	31.9
	One to three times per week	35.4
	Once or twice per month	17.2
	Once and above per quarter	2.6
	Sporadically or not at all	12.9
How often CSRs receive feedback and coaching on telephone technique and service delivery from a supervisor (*n* = 117)	Daily	14.5
	One to three times per week	43.5
	Once or twice per month	38.5
	Sporadically or not at all	3.4
Purpose of work monitoring		
Substantiating disciplinary actions for CSRs (*n* = 114)	A little or not at all	54.4
	Moderate level	26.3
	A lot or a great deal	19.3
Improving CSRs' performance (*n* = 116)	A little or not at all	9.4
	Moderate level	19.0
	A lot or a great deal	71.6
Identifying training needs of CSRs (*n* = 112)	A little or not at all	8.1
	Moderate level	26.8
	A lot or a great deal	65.2

once or twice per month. As shown in Table 5.9, work monitoring was often used for improving CSRs' performance (71.6 per cent) and identifying their training needs (65.2 per cent), rather than substantiating employees' disciplinary action (19.3 per cent).

The efficiency of work organisation at call centres was measured by CSRs' call-handling time and the daily number of calls handled by them. The average handling time was about 2.5 minutes for inbound calls and almost three minutes for outbound calls. The average number of customers that each CSR handled per day was 105.9 inbound calls and 97.7 outbound calls. Eight per cent of inbound calls per CSR per day were abandoned, while, on average, each CSR failed to contact a customer in 43.4 per cent of outbound calls.

Work discretion of CSRs was measured by a variety of indicators, as shown in Table 5.10. CSRs at the surveyed call centres had substantial discretion over daily tasks, working tools and procedures, work pace or speed and over what to say to a customer. Yet, they could use little discretion in designing or using new technology, scheduling breaks and lunch times, revising work methods, handling customers' requests or problems, and settling customers' complaints. Another indicator of discretion at work was the extent of CSRs' use of scripts. The high use of scripts, identified in Table 5.10, was indicative of CSRs' standardised interaction with customers or little work discretion. When comparing work discretion between

Table 5.10 Work discretion of customer service representatives (%)

	Not at all	A little	Moderate	A lot	A great deal
Daily tasks (*n*=116)	2.6	6.9	31.0	47.4	12.1
Work tools, methods, procedure (*n*=116)	2.6	20.7	39.7	29.3	7.8
Work pace or speed (*n*=116)	3.4	11.2	36.2	41.4	7.8
Conversation with customers (*n*=115)	2.6	14.8	40.0	33.9	8.7
Design and use of new technology (*n*=114)	33.3	26.3	25.4	12.3	2.6
Lunch and break scheduling (*n*=116)	12.9	30.2	31.0	20.7	5.2
Revising work methods (*n*=114)	28.1	31.6	27.2	10.5	2.6
Handling customer requests or problems (*n*=116)	9.5	26.7	40.5	20.7	2.6
Settling customer complaints (*n*=115)	9.6	29.6	33.9	21.7	5.2
Using script for talking to customer (*n*=114)	3.5	23.7	26.3	38.6	7.9

in-house and subcontractor call centres, the former had more CSR discretion for every indicator compared with the latter. This implies that subcontractor call centres exert tighter labour control over CSRs than in-house call centres.

Employee participation in work reorganisation is regarded as a key part of workplace innovation to enhance worker morale and performance (Bae and Kang 2003). It is, however, relatively limited at Korean call centres, as illustrated in Table 5.11. About 70 per cent of call centres reported some use of problem solving groups (that is, quality circle or process improvement teams), yet the extent of CSRs actually participating in these groups was only 23.8 per cent, with a wide variation (between 0.3 and 100 per cent) across worksites. Similarly, around two-thirds of call centres implemented a formal suggestion system for CSRs and 35.5 per cent of CSRs in those call centres took part in the suggestion programme. The penetration rate of other workplace innovation practices, in terms of the actual percentage of CSRs participating in these innovative programmes, was between 14.4 and 32.4 per cent – 14.4 per cent for flexible work arrangements, including job sharing, telecommuting and flexi-time, 25.8 per cent for flexible job descriptions of multi-tasking, and 32.4 per cent for job rotations. As such, the limited adoption of workplace innovation implies the dominance of traditional bureaucratic supervision and Taylorised work design over service staff at Korean call centres. It should be noted that almost 70 per cent of call centres had a formal mechanism for gathering customer feed-

Table 5.11 Work reorganisation by customer service representatives (%)

Average percentage of CSRs in quality circle or process-improvement teams (*n* = 103)	23.8
Average percentage of CSRs in flexible work arrangements (*n* = 105)	14.4
Average percentage of CSRs having flexible job descriptions (*n* = 104)	25.8
Average percentage of CSRs in job rotations (*n* = 105)	32.4
Percentage of call centres having a formal suggestion system for CSRs (*n* = 116)	65.5
Average percentage of CSRs participating in the suggestion programmes (*n* = 68)	35.5

Note
Flexible work arrangements include job sharing, telecommuting and flexi-time.

Table 5.12 Call centre technologies in use (%)

Average percentage of daily customer calls completed by a VRU or IVR (*n* = 102)	41.4
Percentage of call centres regularly using the following technologies to interact with customers	
Email (*n* = 118)	72.9
Fax (*n* = 118)	74.6
Media blending (*n* = 118)	27.1
Speech recognition (*n* = 117)	8.5
Workflow management (*n* = 118)	88.1
Electronic customer relationship management (*n* = 118)	27.1
Voice over IP (*n* = 118)	22.0
Web-enablement (joint browsing, chat, instant messaging) (*n* = 118)	39.0

Note
VRU and IVR respectively denote voice recognition unit and interactive voice response unit.

back on CSRs' call service and also performance was appraised through a regular customer survey. Over 50 per cent of call centres provided customer satisfaction data for CSRs at least once per month. This suggests that management at Korean call centres lays particular stress on the high quality of customer service under the control-driven work context.

Recent developments in call centre technologies have produced new opportunities for improving customer interactions by moving from voice-only or telephony channels to multi-media channels, such as email, fax, Internet and voice over Internet protocol. This also enhances the efficiency of the work flow and job assignments. The Korean call centres surveyed introduced some advanced technologies, as shown in Table 5.12. Over 40 per cent of daily customer calls were completed by a voice recognition unit (VRU) or interactive voice response (IVR) without human interaction. New technologies adopted for improving CSRs' interaction with customers were: email (72.9 per cent), fax (74.6 per cent) and work-flow management (88.1 per cent). In addition, 39 per cent of call centres used Web-enablement, including joint browsing, chatting and instant messaging, and 27.1 per cent used media blending as well as electronic customer relationship management (ECRM), 22.0 per cent used voice of Internet protocol (VoIP) and 8.5 per cent used interactive voice recognition. Note that most of the new technologies (except ECRM and Web-enablement) were used more often by in-house operations than by subcontractors.

The environmental context of call centre operations

The majority (87.0 per cent) of call centres were located in Seoul and suburban areas, which reflects the concentration of national economic activities around the capital region. Call centres reported that the location of call centres was mainly influenced by the presence of a skilled workforce (28.3 per cent), key customers (15.9 per cent), better infrastructure (that is, shops, schools, power, transport; 14.2 per cent) and the location of a corporate head office[3] or client companies. However, it should be also noted that 47.1 per cent of call centres, located in the Seoul area, had plans to move to other regions in pursuit of cheaper labour and real estate.

From Table 5.13 it can be identified that the extent of local or regional public resources supportive of call centre operations was minimal. The call centre managers in the survey reported that they made little use of the public resources subsidised by local governments, such as job placement services, job training programmes, site location assistance, regional aid incentives, tax abatements and special loans or grants. This means that local government policies to assist this job-creating sector are either insufficient or not properly targeted. For instance, only 7.0 per cent of CSRs at the surveyed call centres had participated in government-sponsored training programmes. Moreover, management in the survey made written suggestions for the government's active policies to implement job training and placement programmes for call centre employees, provide financial subsidies for call centres' initial investment, regulate client companies' unfair business practices with subcontractors and protect female service representatives from customers' sexual abuse.

Table 5.13 demonstrates that call centres were not networking within their local areas. Participation in regional call centre networking groups was 12.1 per cent, call centre employer association participation was 12.9 per cent and membership of local industrial chambers was 3.6 per cent. In contrast, outside consultants were heavily used with two-thirds employing consultants for training programmes, whereby nearly half used consultants for quality management programmes and technological adoption or re-engineering. These data reveal that call centres were not engaging with local networks or utilising government subsidies but were heavy users of outside consultants.

Table 5.13 Environment of call centres

		%
Reason for the current location of call centres ($n=113$)	Presence of skilled workforce	28.3
	Presence of customers important to call centre business	15.9
	Low wages	1.8
	Low real estate cost	8.8
	Infrastructure (e.g. shops, schools, power, transport)	14.2
	Others (e.g. part of head-quarter office . . .)	31.0
Extent of local or regional public resources supportive of call centre operation		
Job recruitment and placement services ($n=98$)	A lot or a great deal	7.1
	Moderate level	5.1
	A little or not at all	87.8
Training resources or programmes ($N=98$)	A lot or a great deal	5.0
	Moderate level	12.9
	A little or not at all	82.2
Site location assistance ($n=98$)	A lot or a great deal	2.2
	Moderate level	4.2
	A little or not at all	93.7
Incentives for locating in targeted zones ($n=98$)	A lot or a great deal	1.1
	Moderate level	3.2
	A little or not at all	95.7
Tax abatements ($n=98$)	A lot or a great deal	1.1
	Moderate level	3.2
	A little or not at all	95.7
Special loans and/or grants ($n=98$)	A lot or a great deal	0.0
	Moderate level	2.2
	A little or not at all	97.8
Effect of other call centres in the surrounding location ($n=112$)	Valuable resource for recruiting qualified CSRs	10.7
	Constraints to recruitment and retention of qualified CSRs	5.4
	Little effect on recruitment or retention efforts	83.9
Percentage of call centres participating in outside organisations or networks	Local call centre networking group ($N=112$)	12.1
	Employer or trade association for call centres ($n=112$)	12.9
	Regional cross-industry association or chamber ($n=112$)	3.6
Percentage of call centres using outside consultants	Training programmes ($n=110$)	66.4
	Quality management ($n=106$)	44.3
	Technology adoption or re-engineering ($n=101$)	43.6

Conclusion

The key findings from this national survey of Korean call centres are as follows. First, like many Western countries, Korea has witnessed the rapid growth of the call centre sector which applies information and communications technologies to the customer service operations of various industries. This is a very nascent industry for Korea, as almost 60 per cent of call centres have been established since 2000. As a result, call centres have become an important services and employment generating sector.

Second, the employment relations of Korean call centres are substantially externalised, in that a number of call centres (27.6 per cent) are outsourced to subcontractors and over 90 per cent of CSRs of the remaining in-house call centres are non-regular employees. The externalised employment relations at call centres are not only associated with the feminisation of the customer service workforce, but also lead to the segmentation of job career ladders between non-regular or subcontractor CSRs and regular (managerial) employees. This also contributes to union-free labour–management relations at Korean call centres. This is similar to the findings for the US call centre industry (see Srivastava and Theodore this volume), but different from the Swedish (Lindgren and Sederblad this volume) and German (Weinkopf this volume) call centre sectors.

Third, Korean call centres are characterised by hybrid work organisation generated as a result of management's dual interests in cost saving and service quality (or customer satisfaction). In the hybrid work system, intense real-time work monitoring is imposed upon CSRs, while some degree of work discretion is also given to them. However, the hybrid work organisation of Korean call centres comes closer to a supervisor-driven electronic bureaucratic model, in light of the CSRs' limited participation in workplace innovation projects (Lee *et al.* 2004).

Fourth, the study indicates that very limited public resources and sector/local-level organisational networks exist to support the development of the call centre sector. This contrasts with the situation in the UK and Australia. The proliferation of call centres – whether in-house or subcontracted – has been led by corporate management's strategic drive to enhance customer relations by using advanced information and communications technologies. However, despite the call centre sector's great potential for creating jobs (especially for the female workforce), the Korean government has made little effort to foster this sector, unlike some Western countries (such as the UK and Australia) and in developing countries (such as India) where the government is an active player in attracting call centre operations for job creation.

This study provides an overview of the key features of employment relations and labour issues in Korean call centres, by drawing upon national survey data. The next stage of research involves a comparison of business models and employment relations practices between a late-developing country (Korea) and advanced Western countries through utilisation of the global call centre survey database.

Notes

1 This accounts for about 1.5 per cent of the total active workforce in the country (Korea Labor Institute 2005).
2 Call centres in this survey report that 10.4 per cent of CSRs are college students on the active list.
3 This survey finds that corporate head office policies also tend to have a substantial influence on the adoption of job redesign and advanced HRM practices by call centres.

References

Bae, K. and Kang, H. (2003) 'New Labor Relations in the Knowledge-Based Industry: a Case of Mobile Telecommunications Service Industry', Seoul: Korea Labor Institute (in Korean).

Batt, R., Doellgast, V. and Kwon, H. (2004) 'The U.S. Call Centre Industry 2004: National Benchmarking Report: Strategy – HR Practices, and Performance', Report of the Global Call Centre Industry Project.

Belt, V. (2002) 'Women, Social Skill and Interactive Service Work in Telephone Call Centres', *New Technology, Work and Employment*, 17(1): 20–34.

Buchanan, R. and Kock-Schulte, S. (2000) 'Gender on the Line: Technology, Restructuring and the Reorganisation of Work in the Call Centre Industry', Ottawa: Status of Women Canada.

Callcenter Information Research Centre (2004) *Basic Statistics of Korean Call Centre Industry* (in Korean).

Frenkel, S., Korczynski, M., Shire, K. and Tam, M. (1999) *On the Front Line: Organisation of Work in the Information Economy*, Ithaca: Cornell Press.

Holtgrewe, U., Kerst, C. and Shire, K. (2002) *Re-organising Service Work: Call centres in Germany and Britain*, Burlington: Ashgate.

Houlihan, M. (2002) 'Tensions and Variations in Call Centre Management Strategies', *Human Resource Management Journal*, 12(4): 67–85.

Korea Labor Institute (2005) *2005 KLI Labor Statistics*, Seoul: KLI (in Korean).

Lee, B., Kang, H. and Kwon, H. (2004) 'Diverging Convergence in the ICT-driven Employment Relations: a Comparative Case Study of Korean Call Centres', Paper presented to the *2004 Asia-Pacific IIRA Conference*, Seoul, 23–25 June.

Lee, B., Kang, H., Kwon, H. and Kim J. (2005) *Employment Relations and Labor Issues in the Korean Call Centre Industry*, Seoul: Korea Labor Institute (in Korean).

Mulholland, K. (2002) 'Gender, Emotional Labour and Teamworking in a Call Centre', *Personal Review*, 131(3): 283–303.

Taylor, P. and Bain, P. (1999) 'An "Assembly Line in the Head": the Call Centre Labour Process', *Industrial Relations Journal*, 30(2): 101–117.

6   Skill and info-service work in Australian call centres

Bob Russell

Introduction: the political economy of call centres

Call centres have appeared sphinx like, on the terrain of the new economy to assume a dominant role in contemporary service delivery. There are a variety of reasons for this, some of which are 'drivers', while others are clearly 'enablers'. Amongst the former, the prolonged growth and demands placed upon the service economy loom large. Australia, like other OECD countries, is now principally a service-oriented economy, with three-quarters of the workforce employed in a variety of business, personal and public services (ABS 2004a). Indeed, the service sector, broadly defined, has served as the great employment sponge over the last 30 years, absorbing new entrants into the labour market (women and youth) as automation and offshoring have taken their toll on employment levels in the primary and secondary sector manufacturing industries (Castells 1996; Rifkin 1995).

The production and delivery of services depend mainly upon the human use, manipulation and actioning of information, which can vary from handling simple, straightforward requests with minimum amounts of information exchange to more abstract and subtle forms of interpretation and analysis. Information, both verbal and textual, and human labour are the main inputs referred to in this chapter as *info-service* work activity, a term that is explored in greater detail below as being particularly appropriate for the study of call centres.

The extension of the info-service economy at first relied upon massive infusions of labour into its ranks. Productivity tended to remain dormant or increase only haphazardly in these pursuits, despite the concentration and pooling of labour in large office complexes (Brenner 2003; Greenbaum 1995). However, info-service work was subject to the same pressures that exist elsewhere in the economy, including investors' expectations in the private sector and the finite nature of resources, budgets and fiscal constraints in public and third sector undertakings. Such pressures were to give way first to the waves of corporate re-engineering, and privatisation, followed by the popularity of business process outsourcing (BPO) and public–private partnerships that became commonplace after the early 1990s (Benner 2002; Hammer and Champy 1993; Shields and Evans 1998).

If economising pressures were impelling a search for new forms of organisation and new divisions of labour, new mediums for carrying, storing, distributing and manipulating information rendered such reorganisations a real possibility. Workstation computing power and dense channels of distribution, combined with revamped technologies (telephony) to constitute what would collectively become known as the information technologies and these served as powerful enablers for economic change (Castells 2001; Bell 2001). Built as they were upon low value physical inputs, new knowledge-based engineering and design industries use knowledge inputs to process information in order to create new knowledge-intensive products. Such technologies are both flexible and portable. Being less spatially dedicated than the technologies associated with mass industrial production, they also hastened and contributed to the new era of global production and competition (Dicken 2003).

As others have noted, competition comes in various guises, incorporating different strategies (Porter 1985). Quality and reputation, as well as cost competitiveness, are both important vectors of competitive strategy. For many, including proponents of total quality management (TQM), quality and cost are no longer trade-offs. In the end, it is costlier to produce and grapple with poor quality than it is to produce customer satisfaction (Deming 2000). Competitive strategy, which has assumed a new importance in an IT-enhanced, globalising world, has had to take on board both of these factors – efficiency and quality – in a simultaneous fashion (Korczynski 2002). This, however, presents specific challenges and contradictions, which may be more easily glossed over, or at least delayed, in manufacturing than in info-service work environments.[1]

Each of the aforementioned factors – the quest for productivity improvement in the service sector, the availability of new information technologies and the growth of multi-dimensional forms of global competition – has promoted the development of customer contact centres as a format for the rationalisation of info-service work. Thus, call centres usually entail a concentration of labour. As in the financial sector, although by no means restricted to it, branch offices and face-to-face encounters are replaced by over-the-phone, voice-to-voice encounters, which can be more tightly scripted, and controlled through standardised training in designated processes. Smaller, over-the-counter offices and service centres tend to give way to larger call centres in the process.

Economising on info-service labour takes place in various ways. Concentration of work activity is conducive to savings in rents and property taxes, as is the spatial flexibility and locational options that are presented by call centre organisation (Burgess and Connell 2004). Furthermore, the clustering of such work in spatial nodes encourages the development of specialised labour pools, as well as an economising on, and in some cases externalisation, of training costs. Concentration of employment in larger facilities also brings with it possibilities for tighter surveillance and control over labour through the automation of certain aspects of the work, such as the delivery of customers to workers via automated call distribution queues and interactive voice recognition technologies. This equates to greater numbers of transactions per hour and per day. The adoption of such technologies, while encouraging and hastening employment concentration, also favours a tighter regulation over the content of info-service work, or what now passes for quality control.

Finally, a relative concentration of employment in individual worksites can also coincide with the dispersal of employment across global urban and rural settings to take advantage of lower labour costs in particular locales. Such diffusion was first noticeable with the establishment of new greenfield sites within particular regions of national economies. This occurred in developments in northern England, Scotland and Ireland, as well as in rural areas of America, Canada and Australia (Nebraska, New Brunswick and Tasmania) (Richardson *et al.* 2000; Richardson and Belt 2001). Recently, the net has been cast much wider, with the outsourcing of customer contact work to cities in India, (Bangalore, Delhi), the Philippines (Manila) and other impoverished areas, where labour costs are a fraction of those pertaining in Australia (As-Saber *et al.* 2004; Callcentres.net 2004; Taylor and Bain 2004, and in this volume).[2]

Placing these tendencies in context, while drawing an historical analogy, the precedent that would most likely come to mind is the transition from cottage or handicraft forms of material production to factories and finally to mechanised factory regimes (Marx 1971). In this case, the transition would cover the movement from small multi-functional offices to large back office operations and finally to the contact centre, which combines both front and back office tasks. Like many analogies, however, this one can be overdone and is imperfect. A major issue in the analysis of info-service work is exactly how closely such activity emulates or diverges from the industrial model of work and job design. In other words, what types of jobs *are* being generated and what type of jobs *can* be created in info-service work? Is there a difference and, if so, what are its determinants? These issues form the central concerns of this chapter, with the discussion being largely confined to call centre development in Australia.

Call centres in Australia

The issues discussed in this chapter thus far are beginning to be addressed in Australia and elsewhere and have been prompted by two factors. First, the overall growth in info-service labour remains impressive, regardless of the particular indicators that are examined. For instance, the number of call centres (defined as operations with five or more seats and utilising computer-driven call distribution systems) has grown from an estimated 550 sites in 1998 to 4,000 in 2003 (Budde 2004). Employment wise this represents a growth rate of close to 40 per cent over the six-year period, or an increase of 60,000 jobs, from 160,000 to 220,000. To put this in another context, more people now work in Australian call centres than in each of the following industries: mining; textiles, apparel and footwear; printing, publishing and the media; food and beverage production; and finally the manufacturing of metal products including cars (ABS 2004b).

Another measure of the importance of this form of organisation is the actual amounts of business that are transacted through contact centres. It is estimated that in 2003, 75 per cent of all customer contacts occurred through call centre venues (Budde 2004). Typical of these trends is one large public sector organisation that this author has studied. In this case 76.6 per cent of its public contacts were handled through its call centre (1.7 million calls per year), with the remainder split between face-to-face (22.5 per cent) or email (0.8 per cent) contacts. Both inbound and outbound centres in Australia are still largely dedicated to service provision (36 per cent of inbound activity and 43 per cent of outbound), although selling and related functions make up an increasing share of call centre business. In short, with regard to large segments of the service sector (banking and finance, telecommunications, IT services, business services), customer contact centres have become the preferred means of conducting relations with the market, just as more traditional undertakings such as manufacturing industry are increasingly likely to use this form of market mediation.

If anything, contact centre employment growth has been even more rapid in several poorer countries, including India, the Philippines, Malaysia and South Africa. Work that entails the use of highly mobile, flexible technologies, and has a labour bill that routinely constitutes between 60 and 65 per cent of total costs, as is the case in Australia (Budde 2004), is an attractive candidate for global outsourcing (see Srivastava and Theodore this volume). Given this possibility, various interests, including the Australian Council of Trade Unions, as well as state policy makers, have speculated upon the possibilities of positioning Australian jobs within a new global division of info-service labour. Under this scenario, 'long-term, well-paid, skilled employment opportunities' that entail training, occupational development and further career pathways into e-commerce and customer/enterprise relationship management (ACTU n.d. a) would stay onshore, while presumably the less skilled routine service components of info-service work would be prime candidates for overseas' outsourcing.

In effect, such proposals advocate the adoption of a 'high road' strategy for the domestic info-service economy (Batt 2000; Kinnie *et al.* 2000a; Russell 2004a). This would consist of commitment to training and skill development, a focus on quality service delivery, appropriate salary levels and conditions, access to union representation and full consultation around the setting of targets, performance monitoring and other issues affecting the quality of work life (ACTU n.d. b). While this is a laudable vision, at least as far as domestic developments are concerned, it still begs the question: are these the type of jobs that are likely to be created in the info-service economy?

Three different responses to this issue can be identified in the existing call centre literature. First, there are those who argue that info-service work is bound to reproduce the now familiar models of industrial job design, work organisation and divisions of labour. Then there is a body of research that suggests info-service work is a harbinger of a brighter post-industrial future. Finally, a third alternative argues that info-service employment throws up new contradictions that are both unique to this form of labour and that are not anticipated in the knowledge economy literature. This discussion requires that the concept of info-service work and its implications for skill and working conditions be elaborated upon and this is undertaken in the following two sections. Following that, labour force survey results that include four Australian sites operating in different sectors of the economy are analysed for the insights they offer on skill requirements and associated opportunities in the call centre.

Opportunity, oppression and contradiction in the call centre

The popularity and growth of computer-enabled customer contact centres has spawned a number of powerful images that seek to capture the essence of this form of work organisation. 'Big brother' institutions of electronic surveillance (Fernie and Metcalfe 1998), electronic assembly-lines (Taylor and Bain 1999), battery farms (Crome 1998), customer-oriented or mass-customised information centres (Frenkel *et al.* 1999) and quasi-professional, high involvement work systems (Batt 1999, 2000, 2002) are all metaphors that have been deployed in the analysis of call centres. Such images are important as they depict very different futures with respect to training, recognition, remuneration and opportunity for the workers who populate such sites.

Simplifying matters somewhat, it is possible to distil three broad stylisations that may be associated with the growth of info-service labour. The first would see this work firmly located within industrial modes of employment. This is the 'assembly line in the head' variant, where conformity to quantitative key performance indicators inevitably leads to work disaggregation, intensification and process standardisation – the hallmarks of scientific management (Bain and Taylor 2000; Taylor and Bain 1999; Collin-Jacques 2004). Thus, while noting the variation that exists between individual workplaces, in their overview of Scottish call centres, Bain and Taylor flag the predominance of quantity-driven operations, with repetitive, routinised and stressful conditions (Bain *et al.* 2002; Bain and Taylor 2002; Taylor and Bain 2001). The gendering of info-service work, the specific tasks that women perform within call centres and what is recognised as skill as opposed to naturalised talent (Callaghan and Thompson 2002; Belt *et al.* 2000, 2002; Belt 2004; Jenson 1989) also follows well-established patterns established in both older service industries (Hockschild 1983; Martin 1991) and manufacturing undertakings (Parr 1990).

In socio-technical terms (Trist *et al.* 1963), call centres are defined and driven by the operant technologies, to which CSRs are required to adapt. Automated distribution systems and the data gathering and monitoring capacities that are embedded within them limit employee autonomy with respect to both the pace and content of the work (Callaghan and Thompson 2001). This implies a proliferation of comparatively low skilled service jobs. Accordingly, the focus of info-service work is clearly on the latter – service – aspect of the couplet, according to those who underscore the industrial dimensions of the work. This is accompanied by many of the maladies of such employment, including high levels of stress, 'burnout', 'churn' and general job dissatisfaction (Deery, Iverson and Walsh 2002).

While such outcomes are a possibility, they are not inevitable according to other researchers. Indeed, the logic of service may lead in other directions that entail modifications of standardised, Tayloristic job designs, allowing for more scope and discretion for employees to carry out their jobs (Gutek 1995). Batt (2000), for example, suggests that when business strategy turns towards 'relationship management' as opposed to short transactions, mass production models will be sub-optimal. Quasi-experimental evidence suggests that they will be outperformed by CSRs who are organised in self-managing work teams (Batt 1999; Batt and Moynihan 2004a). Thus, there are choices to be made when it comes to designing call centre work (Collin-Jacques 2004). High performance work systems (HPWS) are more likely to be used when business strategies privilege choice and customisation and when the market is willing to pay for such service. They entail a greater focus on training and coaching, team-centred work, group incentives and internal career structures (Batt 2000; Kinnie *et al.* 2000a). This leads to an enhancement of skill levels, including greater reported control over the choice of tasks, tools, procedures, work pace and the timing of work breaks, which has led some to refer to a quasi-professionalisation (Frenkel *et al.* 1999) of these occupations. This movement is given an added impetus when service and sales functions are combined together in one job.

Although HPWS designs are most frequently encountered in business service delivery (Batt 2000; Kinnie *et al.* 2000b), where relationship management is expected to have the greatest payoffs, it need not be restricted to this end of the market. Customisation is likely to seep further downwards into the household sector, becoming a normalised expectation,

although the implications of this for introducing new levels of social strati-
fication (user pay for sharply different grades of service) have largely been
ignored in these analyses. Multi-skilling, in which agents are able to
provide service to a number of different market sectors (household, business) depending upon demand, is another possibility. Finally, the tendency
to combine service and sales roles in each position is likely to entail skill
enhancement. In each instance, the possibilities for skilled, quality job creation are there to be realised (Herzenberg *et al.* 1998).

The application of socio-technical theory to call centres in the HPWS
literature raises a number of important issues. Work designs may be
selected and they may give precedence to informational skills over service
functions. In this case, technology is used to enable rather than to control
and quality service is traceable back to skilled, knowledge workers. A
growing emphasis on quality, such as is found in TQM-inspired discourses,
may then auger well for skill enhancement in customer contact work, especially if this is related to superior performance. By implication then, at
least some sectors of customer service work are prime candidates for
employment upgrading.

This assumes, however, that criteria other than economising efficiency
will be brought to bear on job design. Apart from some public and third-
sector activity, the scope for challenging this logic through more humanis-
tic job designs is probably very limited. Thus, from the episodic evidence
that is available, there seem to be as many examples of organisations
moving away from limited employee autonomy to greater levels of control
as in the opposite direction (Batt 1999; Russell 2004b). Furthermore,
quantification and standardisation is invited by the technology itself. As
call centre technology assumes largely invariant forms, its role in structur-
ing work processes should not be underestimated. As much of the data on
HPWS is provided through managerial surveys it may be prone to under-
estimating the constraining role of current technologies while lending
greater significance to work teams and the other trappings of HPWS than
is actually merited. Managers may be disposed to overemphasise the positive and unique features of their organisations. Given the technological
infrastructure of the call centre, it may well be that workers experience the
attributes of HPWS in a less enthusiastic light than either managers or
researchers. This is not to imply that skill enhancement and greater job
discretion are not positive goods, or that technologies have to be designed
to control their users. Rather, it is to suggest that HPWS may not have
sunk very deep roots into the info-service work environment. Given the
common technologies of the call centre, it remains an open question as to
whether the constituents of HPWS are more superficial than real for those
who do the work.

Somewhere between the 'trapped in the past' and the post-industrial
possibilities of the near future is a third alternative, which emphasises the
unique contradictions that are produced in the call centre. This follows

from the presence of competing demands and the logics they give rise to. On the one hand, service *delivery* is subject to the same norms as any other production process in a capitalist market economy. Cost minimisation, product (in this case service/sales) turnover and value accretion favour production regimes of high managerial control. This includes a prescriptive division of labour, process standardisation such as scripting and tight supervision in the form of monitoring. But alongside these features are the novel aspects of over-the-phone, computer-aided service work, which give rise to other expectations. Info-service work entails both the production/retrieval and dissemination of information and the immediacy of working on (behalf of) clients. It combines production and service in one instant. It also brings a third party, a public, directly into the process. This gives vent to a number of contradictions that revolve around the production of info-service labour.

Employers bring efficiency norms to the table, while clients bring service expectations that can only partially be fulfilled through the former. The CSR is charged with mediating these dual expectations or logics (Korczynski 2001), with or without the autonomy that may be required to accomplish this feat. In more familiar terms, this is what lies at the back of the competing demands for meeting both quantifiable production and qualitative service targets. One means to this end is through the provision of 'rationalised emotional labour' (Hockschild 1983; Korczynski 2001), another contradiction in terms, which CSRs confront on a daily basis. This entails both self-management of one's emotions and management of 'the other' (i.e. client). But such management takes place against the backdrop of rising expectations on the part of the public, encouraged by contemporary discourses that privilege the consumer and her rights. This leads Korczynski (2002) to refer to the 'pleasures and pain' of doing customer service work, or what some refer to as the phenomenon of 'loving the work and hating the job'. In this dynamic the satisfaction of providing helpful service confronts the realities of process rationalisation on the one hand, and the rising expectations of the consumer society on the other. At the micro-level this is a potentially explosive situation that is manifested in public disenchantment with unfulfilled service expectations, employee dissatisfaction, recrimination and abuse. Thus, added to the frictions between employer and worker is another dimension, of working both *for* and *on* the customer, 'our friend, the enemy', a source of both pleasure and pain for info-service workers. It is no wonder then that behind the impressive socio-technical system of the call centre lies a fragile social order that encompasses potentially conflicting demands and expectations.

This rich complexity of info-service work is not fully captured in either industrial analogies or post-industrial HPWS approaches. Rather, info-service labour calls for both discretion in order to satisfy client needs and process standardisation in order to fulfil the budgetary targets of employing organisations. The former requires more highly skilled autonomous workers, while the latter implies a deskilling logic. In the remainder of the chapter these competing dynamics are analysed in the context of four quite different customer contact centres.

Skill in info-service work

Given that call centres first and foremost represent a *process* for the production, treatment and delivery of information, and only, in some instances (e.g. outsourcing) an industry in its own right, a range of skill expectations may be in effect across different centres. Useful for the purposes of understanding skill in this type of work environment is the term that has been employed so far in this chapter, namely info-service labour. Taking this concept literally provides us with the two main dimensions of this type of employment – information work and service delivery. In part, the skill *demands* of client contact work depend upon how information work and service labour are combined into one job function. The former entails the collection/retrieval and interpretation of information to create results for a client/user. The product provided by the CSR invariably includes information that is provided to the caller. It may also include a tangible service, such as processing an insurance claim or sending a tradesperson to a residence. Finally, info-service work may entail the production of additional new information on a public and its relationship with the employing organisation. This is usually stored for future encounters. In the process, information is passed back and forth between CSR and client. Over each job cycle, information obtained through past or current interactions is acted upon to create outcomes that will also involve a new information component. Thus, production and consumption are partly overlapping and partially distinct as signified by the division of the job cycle into different segments such as 'prep time', 'talk time' and 'wrap time'.

As in the case of other service functions, there are also intangible outcomes; information is partially consumed in the over-the-phone encounter, and of necessity it may be tailored to varying degrees to meet the needs of the individual client. In these respects, a *service function* is omnipresent in all call centre work. When the service and servitude dimension of the work is primary, the flow of information is principally from the customer to the CSR for actioning: 'please give me my current bank balance'; 'book me a one-way ticket on Friday ...'; 'discontinue my phone service beginning on ...'; etc. On the other hand, when information exchanges are more complex, involving two-way flows, greater interpretive labour including interaction with both clients and databases, as well as the processing of information, may demand greater skill utilisation. This will often be reflected in a longer job cycle as well as greater agent discretion and less process standardisation.

Job complexity in the call centre can then be viewed as a function of a number of factors. The intricacy or amount of information that inheres to the purchase and/or use of a product, service or public good is one consideration. For example, the amounts of information that are required to purchase and 'consume' an airline ticket will likely be of a different order of magnitude than that which is required to debug a malfunctioning software program. Increasingly, goods and services that have a small information component are the subject of self-servicing through on-line web 'encounters' and interactive voice technologies that don't necessarily involve a human agent. More complex services that contain a high informa-tion component and greater amounts of interpretive effort will continue to involve the labour of intermediaries in contact centres. Product/service complexity, in turn, is partly a function of how producers choose to bundle what they offer and the legal conditions that are attached to operation. Providers may be torn between a customisation strategy of multiple pack-ages and options on the one hand with a higher informational content, and the economies of standardised processes and a lower informational order-ing, on the other hand. With respect to the legal environment of info-service provision, recent provider disclosure laws, as well as 'duty of care' considerations, are likely to add to the informational content of provision.

Up to this point, it is mainly the skill demands of call centre work that have been considered. The other aspect that must be taken into account when assessing the skill profile of info-service work is the qualifications and expectations that workers bring with them to the job. This takes us into the realm of the human resource management function and specifi-cally recruitment. How closely do managers attempt to match qualifica-tions with the skill demands of the work? Prima facie, it would seem as though the matching of individuals to the job was a major consideration, with many centres employing multi-stage interviewing and vetting proce-dures, prior to substantial investments in training (Callaghan and Thomp-son 2002; Thompson *et al.* 2004). Yet quitting and exit rates speak to a different reality, where the fit between job expectations and job character-istics is less than ideal.

This raises different possibilities. One is that HR managers adopt what has been termed a sacrificial strategy (Wallace *et al.* 2000), but which is probably better understood as *cynical* HR. Here, a self-replenishing labour pool, often of young people, is continually available, while training and skill sets are kept to a minimum. This is the 'Mcdonaldisation' variant (Ritzer 2000) of info-service work. The expectation on the part of man-agers and workers is that job attachment will be instrumental and short term. As a strategy, this would only align with call centres that are at the service rather than informational end of the spectrum, as well as with a flush labour market. It would not auger well with more complex interac-tional work, which is likely to place more stringent demands upon local labour markets.[3] A more interesting issue concerns the question of recruit-ment when there is a mix of informational and service features, a problem to which we return in the following sections.

The call centre study

In this section results pertaining to the skill demands of info-service work in call centres are presented. The findings are based upon a workforce survey conducted at four contact centres in the financial services, utilities, state licensing and transportation sectors.[4] For ease of reference the research sites are referred to as *Pensions, Powerco, Government services* and *Destinations*, respectively. A total of 220 surveys were completed from amongst the participating organisations, which ranged in size from 40 CSRs to over 150. While hardly definitive of all call centres in Australia, as described below, these sites are by no means atypical of the call centre work environment. As such the findings offer intriguing previews into how CSRs evaluate the skill requirements of their work. They also let us specify the determinants of such assessments. Thus, an important aim of the study is to determine how agents perceive the skill demands of their work and to analyse factors that may influence such evaluations.

The survey was supplemented by extensive field observation at each centre on the part of the author. Typically, this entailed a series of visits in which the researcher would conduct side-by-side listening with designated members of the workforce, followed by additional debriefing questions. In addition, one or more open-ended interviews were conducted with each call centre manager, as well as with team leaders, and at some centres, the trainers. These sessions provided a sense of context, history and further clarification for the responses that the survey was generating.

Each of the centres employed similar technologies, including automated call distribution systems and visible neon signage boards that provided real-time information on call volumes, queues and the quality of service as measured by waiting times. 'Drop down' call technology had not been adopted at these centres, which meant that agents still had to 'pick up' the call. Thus, it is possible to hold technology constant as a variable, while analysing the impact of other factors in the organisation of work on job skill.

At three of the four centres, CSRs were dedicated to specific products. The transportation centre exhibited the simplest structure, where all 70-odd agents did exactly the same thing in providing the public with information on urban transit schedules. Here, the product was informational only, but it did not entail the retrieval or storage of information on particular users. Typical calls were of a short duration and involved the use of only one software program, a location finder. Consequently, training periods were also atypically short, entailing only one week, followed by another week of side-by-side instruction. Employees held a trainee designation for one month, after which they graduated to part-time, casual status. This centre employed many university students who would typically put in a 12 to 15 hour work week as a CSR.

At the financial services and utilities call centres CSRs worked in designated product teams/areas. In the case of the former, specialisation was by superannuation fund; employees were assigned to different industry funds in either an inbound customer service or outbound customer information role. Training periods here were twice as long as at the transportation centre (two weeks), but ongoing training in new product knowledge and legal requirements was also a feature of this centre. At the time of the study, moves were afoot to establish more responsible work teams that were assigned certain self-managing functions. For example, teams were given control over the rostering of their breaks, while team meetings were chaired by members on a rotating basis, at venues of their choice. In spite of this devolution, no thought was given to abolishing the team leader positions at this company.

Similar structures were in place at the utility company, where, with one exception,[5] agents worked in teams that specialised in either electricity or natural gas service. Intake training at this centre took up to eight weeks and was equally divided between classroom and on-the-job instruction.[6] A notable feature of work at this centre was the high percentage of calls that required negotiation with the residential clients over accounts in arrears and the establishment of repayment schedules. It was estimated that up to half of all calls received involved an element of negotiation, which in turn may have accounted for the longer training periods.

The state licensing call centre differed from the above in one potentially significant feature. Work teams, which had once been assigned to specific functions and services, such as vehicle or marine registration, or personal licences, were now multi-skilled to undertake any of the functions the call centre provided. Standard intake training periods of four weeks were in effect at this centre, combining a mix of classroom and on-the-job buddying. As at the financial centre, heed was also being paid to the creation of greater levels of team self-management. Teams selected their own 'captains', set their own agendas at the fortnightly meetings and at the time of the study were working on a new rostering plan that would be presented to management as a fait accompli.

With the exception of the transportation centre, work at the other locations appeared to be '*multi-layered*'. That is, it involved a range of questions of varying complexity from the centres' respective publics, entailed familiarity with a host of products or services, and utilised different software programs or databases. The two public sector centres did not employ any casual staff, while the financial centre had a mix of permanent and casual employees and the transportation centre consisted almost entirely of a casual workforce. Although each of the centres engaged in call monitoring practices, workers at each of the centres, apart from *Destinations*, could elect the mode of monitoring (silent versus side-by-side) that they were subject to. Apart from *Pensions*, we see little of the market segmentation strategies that have been detected in other operations (Batt 2000; Kinnie *et al.* 2000b), but which may be particularly suitable to the financial sector or other quarters (e.g. Telco's), where it is possible to stratify amongst households and between residential, small business and large business users. At *Pensions*, in addition to teams being assigned to specific industry pension funds, one outbound team was dedicated solely to working with employers to gather information, run campaigns and retain corporate membership. At *Government services* and to a lesser extent at *Powerco*, initiatives in multi-skilling at team level had been undertaken. Elements of self-management were also in effect at *Pensions* and *Government services*.

A summary of the salient differences in the centres is presented in Table 6.1, which shows interesting variation along a number of the features discussed above. In the next section, we examine the effects of these policies on skill development amongst CSRs.

Skill in the call centre

Three questions in the survey related directly to the skill level CSRs considered they were called upon to exercise in their work. Table 6.2 displays the questions and the results by call centre. Regardless of the centre, large proportions of the sample disagree with the statement that their current positions are the most skilled jobs they have ever held. Moreover, at three of the centres approximately two-thirds disagree that their jobs are making use of their education and experience, while at the fourth centre, *Destinations*, the overwhelming majority cite a mismatch between the qualifications they bring to the job and the requirements of the work. As indicated, this is the one finding that is statistically significant. This should not come as a particular surprise, given that this centre employs mainly university students and takes relatively simple queries from the public.

The findings at the other three centres and the discrepancies between people and the jobs they hold are a more startling finding. This may or may not translate into job dissatisfaction. Thus, opinion is more evenly split over satisfaction with the opportunities that the work provides to use skills. Between a third and half of respondents at *Pensions*, *Powerco* and *Government services* are satisfied with opportunities to use their skills. Not surprisingly, survey respondents at *Destinations* exhibit lower levels of satisfaction in this regard as well. As anticipated, the one call centre that combines a comparatively highly educated labour force with relatively simple service functions is an outlier along the skill indicators that are used in this study. However, it is still the case that at the remaining centres the majority disagree that their current jobs are the most skilled ones they have held, or that the jobs are making optimum use of individual capabilities. Outside of *Destinations*, views are more evenly divided on satisfaction with opportunities to use skills in the call centre.

As suggested above, perceptions of the skill content of call centre jobs may be hypothesised as being a function of the relationships between

Table 6.1 Call centre research site characteristics

Centre	Product market	Workforce	Training (intake period) (weeks)	Work team organisation	Multi-skilling	Call handling targets[a]	Employee choice in form of monitoring	Performance pay	Trade union density
Destinations	Urban transportation information	Casual part-time students	2	No	No	Hard targets	No	No	Zero
Pensions	Superannuation administration	Partially casualised; full-time	4	Yes; constricted autonomy	No	Soft targets	Yes	No	20%
Government services	Vehicle/driver licensing	Permanent full-time	4	Yes; constricted autonomy	Yes	Soft targets	Yes	No	78%
Powerco	Electricity and natural gas supply	Permanent full-time/part-time split	8	Yes	Limited to one team	Hard targets	Yes	Yes; optional individual performance agreements	30%

Note
a Soft targets are those which refer to team/group norms, such as having 80 per cent of all calls answered within 40 seconds. Hard targets refer to individual performance indicators such as prescribed availability, talk and wrap-up limits.

Table 6.2 Skill perceptions of CSRs (%) (n = 220)

		Companies				
		Destinations	Pensions	Gov't services	Powerco	Total
'This is the most skilled job I have ever held'	Agree	24.0 (6)	22.9 (8)	30.3 (10)	29.4 (35)	27.8 (59)
	Unsure	12.0 (3)	5.7 (2)	0 (0)	12.6 (15)	9.4 (20)
	Disagree	64.0 (16)	71.4 (25)	69.7 (23)	58.0 (69)	62.7 (133)
'Job makes use of my education and experience'[a]	Agree	0 (0)	31.4 (11)	21.2 (7)	27.1 (32)	23.7 (50)
	Unsure	4.0 (1)	8.6 (3)	12.1 (4)	7.6 (9)	8.1 (17)
	Disagree	96.0 (24)	60.0 (21)	66.7 (22)	65.3 (77)	68.2 (144)
'Satisfied with opportunities to make use of my skills'	Agree	24.0 (6)	50.0 (17)	48.5 (16)	35.0 (41)	38.3 (80)
	Unsure	28.0 (7)	14.7 (5)	12.1 (4)	24.8 (29)	21.5 (45)
	Disagree	48.0 (12)	35.3 (12)	39.4 (13)	40.2 (47)	40.2 (84)

Note
a Cramers V = 0.17, *p = 0.06.

three sets of factors: the characteristics and qualifications that an individual brings to a job; the relationship between the job holder and the socio-technical operating system that she uses, or what is more commonly referred to as the labour process; and the social design of the job, which is often reflected in the HR paradigm the organisation subscribes to.

In the following we first examine the bivariate relationships between these factors and reported skill. These results are presented in Table 6.3. As can be seen, older workers (over the age of 46) and those who have longer employment seniority with the organisation (three or more years) are significantly more likely to say that they are making use of their education and experience in their work and to express satisfaction with the opportunities the work presents for utilising skills. On the other hand, there is also a strong negative correlation between level of education and evaluation of the skill demands of the work, as well as a moderately negative correlation between educational attainment and satisfaction with opportunities afforded by the work. While high school graduates are evenly divided over whether the work makes use of their qualifications and are more satisfied than not with opportunity, this quickly tails off for workers with more education. Previous call centre experience and gender are unrelated to the evaluations CSRs make concerning the skill demands of their work or their satisfaction with the match between qualifications and job requirements, although women are significantly more likely to consider their current positions to be the most skilled jobs that they have held to date (not shown).

Several variables relating to HR policy were examined as part of the study. The literature which promotes high performance work systems is especially enthusiastic about the difference the adoption of such practices can make to employee attitudes and performance. The aspects of HR that were focused upon in this study include aspects of both initial and ongoing training activity, including its frequency and modalities, the utilisation and perceptions of 'coaching' in the employment relation, the scheduling and agendas of work team meetings, the use of job rotation that entails off-phone work, and the regular occurrence and perception of performance competitions in the workplace.

As indicated in Table 6.3 several of the bivariate relationships between making use of education and experience on the job, and satisfaction with the opportunities to make use of skills are correlated with HR practices in the workplace. Although the amount of training that one receives does not have an immediate impact on skill perceptions, the mode of delivery does when it comes to assessing whether the job makes use of one's education and experience (although not satisfaction with opportunity). Clearly, class-room based training that makes use of qualified instructors is related to CSR skill evaluation, as is the ability to voluntarily access additional training. In particular, the quality of additional training displays a strong associ-ation with the first skill measure of making use of education/experience.

Regular participation in team meetings also bears a substantial relation-ship with the use of skill and satisfaction with opportunities to use skills, although the content of such meetings and control of the agenda does not figure into the skill assessments that were made. Participation in special off-line projects and especially those that involve organisational or process changes enhance self-reported skill estimates. Finally, a positive attitude towards organised workplace competitions over key performance indic-ators (KPIs), towards personal KPIs and towards the practice of coaching is related to more affirmative evaluations of the skill indicators used in this study. On the other hand, call monitoring is not related to respondent evaluations on the use of education/experience, while those who have more negative impressions of call monitoring are also more likely to dis-agree that info-service work presents opportunities to make use of their skills.

The list of labour process variables that are related to positive skill assessments by workers in a significant fashion is somewhat shorter. Rou-tinisation, job variety and, lastly, the number of computer programs used on the job are related to CSR judgements as to whether they are utilising

Table 6.3 Bivariate correlations with skill perceptions (Gamma coefficients *N* = 211)

Independent variables	Job makes use of my education	Satisfied with opportunities to make use of my skills
Employee attributes		
Age	0.34**	0.19**
Gender	ns	ns
Education	−0.41****	−0.26***
Time with organisation	0.21**	0.16**
Previous call centre experience	ns	ns
Job attributes		
Work variety	0.34**	0.41***
Routinisation	−0.35**	−0.55****
Choose speed	ns	0.16*
Work intensity	ns	ns
Inadequate staffing	ns	ns
Confronted with difficult customers	ns	ns
Read from a script	ns	ns
Responsible for follow-up work	ns	0.19*
No. of software programs used	0.25**	ns
HR attributes		
Initial training time	ns	ns
Use of designated trainers	0.25*	ns
Frequency of additional training	ns	ns
Use of trainers in additional training	0.55****	ns
Volunteer for additional training	0.22*	0.23**
Attendance at team meetings	0.53***	0.26*
Team meetings provide opportunity for real input	ns	0.27***
Involvement in special off-line projects	0.32**	0.23**
Importance of KPIs	0.25**	0.37****
Positive attitude to performance competitions	0.34**	0.37***
Positive attitude to coaching	0.28*	0.29**
Objects to monitoring	ns	−0.29***

Note
* <0.10; ** <0.05; *** <0.01; **** <0.001; ns – not significant.

their education and training in ways that are commensurate with expectations. If the work is viewed as routine and lacking in variety, skill assessments are correspondingly diminished as evidenced by the negative correlation. Workers who need to use more as compared with fewer software programs view their jobs as more challenging.

As for satisfaction with the opportunities to use one's skill on the job, the amounts of variety and the levels of routinisation still feature as the most important considerations. Software knowledge has little to do with satisfaction, while ability to influence the pace of one's work and responsibility for post-call work and longer job cycles are positively related to satisfaction with opportunities to utilise skill. That only a few variables are significantly related to evaluations as to whether potential is being realised is perhaps indicative of the tight constraints on the work and the limited possibilities that inhere to altering it with existent technological packages. That other labour process variables become more important when workers assess the actual opportunities they have to use their skills in the work, as opposed to whether they are making use of the skills they bring to the job, may speak to the diminished expectations and inurement that come with time spent in the call centre.

It is also worth noting that the labour market intentions of these workers is strongly related to the match or mismatch that is presented between the job, what the worker brings to it, and the expectations that are created once in it. Workers who consider that the job makes use of their education and experience and who are satisfied with the opportunities posed by the work are likely to stay put. Those who think otherwise are likely to move on. Dissatisfaction with ability to use education on the job and with opportunities to make use of skills accounts for the intentions to exit that many workers expressed in the survey.

The question of matching workers to jobs also stands out in the multivariate analysis, which is presented in Table 6.4. In this table, the variables which showed the most promise (i.e. strongest bivariate relationships) with the two indicators of skill are entered into regression equations. As in the preceding table, the first model takes as its indicator of skill the job's demands on the use of one's education and experience. In the second model, satisfaction with the opportunities to use one's skills in the job (as defined by the respondent) is taken as the dependent attribute of skill.

As Table 6.4 suggests, the most important predictor of using education and experience in call centre work is current educational attainment. Those with some university experience or those with higher educational qualifications are significantly more likely to discount their work and its capacity to make full use of their qualifications. At *Destinations*, where a highly educated workforce is matched with the least demanding jobs amongst the different centres in this study, the relationship between human capital and required skill perceptions is the most transparent and unsurprising. It could readily be argued that this centre is an outlier in terms of skill requirements (i.e. relatively low) and the human capital that employees bring to the job (i.e. relatively high). A logical extension of the analysis would omit this site from the model, in order to ascertain whether educational attainment still has a negative impact on skill evaluations at the other centres where the work is more complex. Omitting *Destinations* from the analysis (not shown in Table 6.4) does not alter the relationship between education attainment and abilities to use knowledge on the job. The relationship remains negative and significant for those with some university or higher educational qualifications.

The other variables that have a significant positive impact on abilities to

Table 6.4 Regression equations for skill indicators

Model 1 – Make use of education in my job				
	B	*Std error*	*Beta*	*Sig*
Age (Dummy var)				
18–24	−0.155	0.199	−0.060	0.437
35–45	−0.053	0.181	−0.022	0.770
46+	0.108	0.217	0.037	0.621
Education (Dummy var)				
Some high school	−0.848	0.975	−0.298	0.386
High school grad	−1.436	0.966	−0.572	0.139
Some college	−1.536	1.004	−0.350	0.128
College grad	−1.415	0.968	−0.560	0.146
Some uni	−1.828	0.976	−0.569	0.063
University grad	−2.015	0.981	−0.616	0.041
Some post-grad	−1.846	1.077	−0.239	0.088
Post-grad completed	−2.103	1.192	−1.764	0.80
Use of job trainers	0.047	0.145	0.022	0.748
Received additional training	0.103	0.232	0.032	0.658
Use of job trainers in additional training	0.199	0.167	0.090	0.236
Participation in off-phone projects	0.137	0.167	0.060	0.413
Participation in organisation/process change	0.472	0.262	0.128	0.074
Participation in team meetings	−0.063	0.260	−0.020	0.808
Enjoys work team competitions	0.298	0.154	0.130	0.055
Volunteer to take additional training	0.062	0.077	0.053	0.427
Job is routine	−0.074	0.090	−0.055	0.410
KPIs are important	0.103	0.094	0.072	0.274
Use 1–3 computer programs	−0.233	0.260	−0.103	0.371
Use 4–6 computer programs	−0.145	0.255	−0.062	0.570
Use 7+ computer programs	−0.319	0.264	−0.130	0.228
Work full time	0.295	0.169	0.123	0.083
Casual employment	0.246	0.226	0.079	0.278
Constant	3.154	1.152		0.007

$R^2 = 0.367$ Adjusted $R^2 = 0.273$

use knowledge at work are participation in special projects, enjoyment of performance competitions and participation in full time work. The first requires little explanation. Time spent off the phones working on special assignments is likely to increase a CSR's skill base. It is the call centre's equivalent to job rotation/expansion in industry. The second HR variable, attitude to performance competitions, may offer some insights into why these 'games' have become such a significant feature of call centre culture. The findings indicate that a positive attitude towards such competitions enhances the sense that the job is making use of the skills that one brings to the job. Perhaps success in such competitions legitimates the sense that

Table 6.4 continued

Model 2 – Satisfied with opportunities to use skills in work

	B	Std error	Beta	Sig
Age				
18–24	0.037	0.175	0.017	0.832
35–45	−0.183	0.159	−0.091	0.251
46+	0.158	0.190	0.066	0.046
Education				
Some high school	−0.876	0.853	−1.027	0.306
High school grad	−1.180	0.841	−0.568	0.163
Some college	−1.150	0.878	−0.308	0.192
College grad	−1.181	0.846	−0.565	0.164
Some uni	−1.321	0.852	−0.502	0.123
University grad	−1.445	0.857	−0.530	0.094
Some post-grad	−1.651	0.938	−0.261	0.080
Post-grad completed	−1.603	1.037	−0.180	0.124
Use of job trainers	0.016	0.127	0.009	0.899
Received additional training	0.001	0.202	0.000	0.996
Participation in off-phone projects	0.097	0.150	0.051	0.518
Participation in organisational/ process change	0.217	0.234	0.070	0.356
Participation in team meetings	0.046	0.229	0.018	0.841
Use 1–3 computer programs	−0.055	0.230	−0.029	0.809
Use 4–6 computer programs	−0.053	0.221	−0.028	0.810
Use 7+ computer programs	0.003	0.230	0.001	0.990
Work full time	0.041	0.148	0.021	0.781
Casual employment	0.0289	0.198	0.111	0.147
Object to use of call monitoring	−0.340	0.132	−0.180	0.011
Use of trainers for additional training	−0.151	0.147	−0.083	0.306
Enjoys work team competitions	0.156	0.139	0.082	0.263
Volunteer to take additional training	−0.008	0.068	−0.008	0.909
KPIs are important	0.223	0.084	0.189	0.009
Job is routine	−0.259	0.079	−0.232	0.001
Constant	4.452	1.012		

$R^2 = 0.312$ Adjusted $R^2 = 0.203$

the job is making demands upon one's skills – that one is becoming better at the job.

While the labour process variables seem to have little impact on evaluations as to whether call centre work is making use of the knowledge that one brings to the job, they are more important in accounting for satisfaction with the opportunities the job presents to utilise skills. Thus, in addition to higher levels of educational attainment, which once again exhibits a negative relationship with satisfaction to utilise skill at work, routinisation also accounts for frustration. Perceptions that the job is highly routine – a commonly held view in call centres – diminishes satisfaction with opportunity. Indeed, routinisation overrides specifics of the job such as the number of software programs that one requires in accounting for worker skill evaluation. It is also the case that respondents who object to the call monitoring processes in place also reveal their frustration with the opportunities that are available to make use of one's skills, while those who attach greater importance to their personal KPIs are more likely to demonstrate satisfaction with the opportunities that are available.

Conclusion

Evaluating the skill demands of info-service work is, in part, a function of the benchmark that is utilised. On the one hand, compared with the mechanised work of the industrial era, info-service work requires more skills and is more demanding in its simultaneity. For the most part, these are not single task, linear functions. Although contact centres may take on some of the features of the assembly line, the work entailed in three of the four centres examined in this study was some way removed from semi-skilled assembly work. On the other hand, our results indicate that info-service work in customer contact centres is also some distance removed from the stylisations of the knowledge economy. The adoption of technologies, including information technologies such as web-based process manuals, that standardise and routinise work lead to these results. In the face of such imperatives, HR initiatives such as those associated with high performance work systems are overpowered by the labour process of call centre work. When potential knowledge workers (e.g. highly educated staff) are recruited for such jobs, dissatisfaction with the skill attributes of the work is commonplace.

These findings point to the newly emergent contradictions that are associated with info-service labour. Much call centre work requires the use of interpretive labour in computer-mediated interaction. The managerial penchant for routinising such work raises new issues that were not present in the manufacturing environment. Recruiting highly educated, reflexive staff and then subjecting them to the provision of ever more standardised service levels adds further to the complexity of work in the contemporary world.

Notes

1 It should not be forgotten that most of the new popular managerial initiatives of the last 20 years, including TQM, had manufacturing industry as their point of reference (Korczynski, 2002).
2 Taylor and Bain cite labour cost advantages of moving from the UK to India in

the order of 60 to 65 per cent. At one large call centre employing over 1,000 CSRs across three sites in Bangalore that this author visited, monthly salaries at entry level were posted at 8,000R (2002), which would equate into $296 (Australian). Even with additional commissions, which could total up to an extra 30 per cent, total monthly salaries would equate to $385 or $4,620 per annum. Comparable Australian call centre wages at the time averaged in the mid-$30,000 range.

3 Retention of workers and especially concerns about the poaching of workers by other contact centres points to tightening labour market conditions for info-service workers in certain Australian cities.

4 One of the study organisations had two call centre sites which both participated in this study. Thus, the survey involved four organisations, but five call centres. A complete description of the participating organisations is found in Russell (2004a).

5 One out of the 16 teams at this centre was 'multi-skilled' and switched back and forth between the two products as necessitated by call demand.

6 Efforts to halve these training times through technological change in software systems and greater process standardisation are described in Russell (2005).

References

ACTUa (no date) *On the Line: The Future of Australia's Call Centre Industry*, ACTU Call Centre Unions Group.

ACTUb (no date) *Minimum Standards Code for Call Centres*, ACTU Call Centre Unions Group.

As-Saber, S., Holland, P. and Teicher, J. (2004) 'Call Centres in India: An Eclectic Phenomenon in Global Human Resource Management', *Labour and Industry*, 14(3): 39–57.

ABS (Australian Bureau of Statistics) (2004a) *Yearbook of Australia*, Canberra: ABS, Table 18.9.

ABS (Australian Bureau of Statistics) (2004b) *Labour Force Australia*, Cat. # 6291.0.55.001, Electronic Delivery, Quarterly.

Bain, P. and Taylor, P. (2000) 'Entrapped by the "Electronic Panopticon"? Worker Resistance in the Call Centre', *New Technology, Work and Employment*, 15(1): 2–18.

Bain, P. and Taylor, P. (2002) 'Ringing the Changes? Union Recognition and Organisation in Call Centres in the UK Finance Sector', *Industrial Relations Journal*, 33(3): 246–261.

Bain, P., Watson, A., Mulvey, G., Taylor, P. and Gall, G. (2002) 'Taylorism, Targets and the Pursuit of Quantity and Quality by Call Centre Management', *New Technology, Work and Employment*, 17(3): 170–185.

Batt, R. (1999) 'Work Organisation, Technology and Performance in Customer Service and Sales', *Industrial and Labor Relations Review*, 52(4): 539–564.

Batt, R. (2000) 'Strategic Segmentation in Front-line Services: Matching Customers, Employees and Human Resource Systems', *International Journal of Human Resource Management*, 11(3): 540–561.

Batt, R. (2002) 'Managing Customer Services: Human Resource Practices, Quit Rates and Sales Growth', *Academy of Management Journal*, 45(3): 587–597.

Batt, R. and Moynihan, L. (2004) 'The Viability of Alternative Call Centre Production Models', in S. Deery and N. Kinnie (eds) *Call Centres and Human*

Resource Management: A Cross-National Perspective, Houndmills, Basingstoke: Palgrave.

Bell, D. (2001) *The Future of Technology*, Subang Jaya, Malaysia: Pelanduck Publications.

Belt, V. (2004) 'A Female Ghetto? Women's Careers in Telephone Call Centres', in S. Deery and N. Kinnie (eds) *Call Centres and Human Resource Management: A Cross-National Perspective*, Houndmills, Basingstoke: Palgrave.

Belt, V., Richardson, R. and Webster, J. (2000) 'Women's Work in the Information Economy: The Case of Telephone Call Centres', *Information, Communication and Society*, 3(3): 366–385.

Belt, V., Richardson, R. and Webster, J. (2002) 'Women, Social Skill and Interactive Service Work in Telephone Call Centres', *New Technology, Work and Employment*, 17(1): 20–34.

Benner, C. (2002) *Work in the New Economy*, Oxford: Blackwell.

Brenner, R. (2003) *The Boom and the Bubble*, London: Verso.

Budde, P. (2004) *Australia – Call Centres.doc*, Paul Budde Communication Pty. Ltd.

Burgess, J. and Connell, J. (2004) 'Emerging Developments in Call Centre Research', *Labour and Industry*, 14(3): 1–13.

Callaghan, G. and Thompson, P. (2001) 'Edwards Revisited: Technical Control and Call Centres', *Economic and Industrial Democracy*, 22(1): 13–37.

Callaghan, G. and Thompson, P. (2002) '"We Recruit Attitude": The Selection and Shaping of Routine Call Centre Labour', *Journal of Management Studies*, 39(2): 233–254.

Callcentres.net (2004) The Contact (11 May; 27 July).

Castells, M. (1996) *The Rise of the Network Society*, Oxford: Blackwell Publishers.

Castells, M. (2001) *The Internet Galaxy*, Oxford: Blackwell Publishers.

Collin-Jacques, C. (2004) 'Professionals at Work: A Study of Autonomy and Skill Utilisation in Nurse Call Centres in England and Canada', in S. Deery and N. Kinnie (eds) *Call Centres and Human Resource Management: A Cross-National Perspective*, Houndmills, Basingstoke: Palgrave.

Crome, M. (1998) 'Call Centres: Battery Farming or Free Range?', *Industrial and Commercial Training*, 30(4): 137–141.

Deery, S., Iverson, R. and Walsh, J. (2002) 'Work Relationships in Telephone Call Centres: Understanding Emotional Exhaustion and Employee Withdrawal', *Journal of Management Studies*, 39(4): 471–496.

Deming, W. E. (2000) *Out of the Crisis*, Cambridge Mass: MIT Press.

Dicken, P. (2003) *Global Shift: Reshaping the Global Economic Map in the 21st Century*, London: Sage Publications.

Fernie, S. and Metcalfe, D. (1998) '(Not) Hanging on the Telephone: Payment Systems in the New Sweatshops', Centre for Economic Performance, London School of Economics and Political Science.

Frenkel, S., Korczynski, M., Shire, K. and Tam, M. (1999) *On the Front Line: Organisation of Work in the Information Economy*, Ithaca, NY: Cornell University Press.

Greenbaum, J. (1995) *Windows on the Workplace: Computers, Jobs, and the Organisation of Office Work in the Late Twentieth Century*, New York: Monthly Review Press.

Gutek, B. (1995) *The Dynamics of Service: Reflections on the Changing Nature of Customer/Provider Interactions*, San Francisco: Jossey-Bass.

Hammer, M. and Champy, J. (1993) *Reengineering the Corporation*, New York: Harper.

Herzenberg, S., Alic, J. and Wial, H. (1998) *New Rules for a New Economy: Employment, and Opportunity in Postindustrial America*, Ithaca: Cornell-ILR Press.

Hockschild, A. (1983) *The Managed Heart: Commercialisation of Human Feeling*, Berkeley: University of California Press.

Jenson, J. (1989) 'The Talents of Women, the Skills of Men: Flexible Specialisation and Women', in S. Wood (ed.) *The Transformation of Work?*, London: Unwin Hyman.

Kinnie, N., Hutchinson, S. and Purcell, J. (2000a) '"Fun and surveillance": the Paradox of High Commitment Management in Call Centres', *International Journal of Human Resource Management*, 11(5): 967–985.

Kinnie, N., Purcell, J. and Hutchinson, S. (2000b) 'Modelling HR Practices and Business Strategy in Telephone Call Centres', *Airannz Conference Proceedings*, Vol. 2, Newcastle NSW: University of Newcastle.

Korczynski, M. (2001) 'The Contradictions of Service Work: The Call Centre as Customer-Oriented Bureaucracy', in A. Sturdy, I. Grugulis and H. Willmott (eds) *Customer Service: Empowerment and Entrapment*, Houndmills, Basingstoke: Palgrave.

Korczynski, M. (2002) *Human Resource Management in Service Work*, Houndmills, Basingstoke: Palgrave.

Martin, M. (1991) *'Hello Central?' Gender, Technology and Culture in the Formation of Telephone Systems*, Montreal: McGill-Queen's University Press.

Marx, K. (1971) *Capital*, Vol. 1, Moscow: Progress Publishers.

Parr, J. (1990) *The Gender of Breadwinners*, Toronto: University of Toronto Press.

Porter, M. (1985) *Competitive Advantage*, New York: Free Press.

Richardson, R. and Belt, V. (2001) 'Saved by the Bell? Call Centres and Economic Development in Less Favoured Regions', *Economic and Industrial Democracy*, 22(1): 67–98.

Richardson, R., Belt, V. and Marshall, N. (2000) 'Taking Calls to Newcastle: The Regional Implications of the Growth in Call Centres', *Regional Studies*, 34(4): 357–369.

Rifkin, J. (1995) *The End of Work*, New York: Tarcher Putnam.

Ritzer, G. (2000) *The Mcdonaldisation of Society*, Thousand Oaks, CA: Pine Forge Press.

Russell, B. (2004a) 'Are all Call Centres the Same', *Labour and Industry*, 14(3): 91–109.

Russell, B. (2004b) '*"You Gotta Lie to it": Software Applications and the Management of Tech Change in a Call Centre*', mimeo available from author.

Shields, J. and Evans, B. M. (1998) *Shrinking the State: Globalisation and Public Administration 'Reform'*, Halifax: Fernwood.

Taylor, P. and Bain, P. (1999) '"An Assembly Line in the Head": Work and Employee Relations in the Call Centre', *Industrial Relations Journal*, 30(2): 101–117.

Taylor, P. and Bain, P. (2001) 'Trade Unions, Workers' Rights and the Frontier of Control in UK Call Centres', *Economic and Industrial Democracy*, 22(1): 39–66.

Taylor, P. and Bain, P. (2004) 'Call Centre Offshoring to India: The Revenge of History?', *Labour and Industry*, 14(3): 15–38.

Thompson, P., Callaghan, G. and van den Broek, D. (2004) 'Keeping up Appearances: Recruitment, Skills and Normative Control in Call Centres', in S. Deery and N. Kinnie (eds) *Call Centres and Human Resource Management: A Cross-National Perspective*, Houndmills, Basingstoke: Palgrave.

Trist, E., Higgin, G., Murray, H. and Pollock A. (1963) *Organisational Choice*, London: Tavistock.

Wallace, C., Eagleson, G. and Waldersee, R. (2000) 'The Sacrificial HR Strategy in Call Centres', *International Journal of Service Industry Management*, 11(2): 174–184.

7 Gender, skills and careers in UK call centres

Susan Durbin

Introduction

At the time of this case study research, call centres were in the third generation of development, characterised by a tried and tested technological platform, which supported well developed processes and operating procedures. The primary focus of the case studies reported here was the transfer of simple transactions from the branch network into a telephony environment. In one case there was, however, an emergent emphasis upon relationship building and sales through the adoption of 'talk time', with advisers being expected to spend at least 65 per cent of their day talking to customers. Since then, we have seen a more aggressive outsourcing of transactional activities to low cost global locations, such as India, and a greater emphasis of the 'sales through service' techniques referred to above. In the UK environment, this has been coupled to the adoption of more sophisticated self-service technology, which acts as a gateway to the call centre operators. While one case organisation outsourced a portion of its operation to India, the other opted to remain solely UK-based, its objective being to use every customer contact opportunity to build a relationship with its customers, thereby enhancing opportunities for cross-selling which, in turn, builds loyalty and ultimately, customer advocacy. What started as a local, tactical response to saving costs was a first step in the development of an alternative distribution channel for financial services – which has now taken on a global dimension.

The nature of call centre work in the UK has been approached from a number of perspectives and developed in several stages, traditionally focusing upon the 'production line' (Fernie and Metcalfe 1998) 'empowerment' (Frenkel *et al.* 1999) and 'worker resistance' (Bain and Taylor 2000) models. More recently, focus has shifted to call centre working in terms of human resource issues (Deery and Kinnie 2004). The outsourcing of UK call centres is emerging as the next stage in the debate, with India dominating global offshore call centre outsourcing by offering Western firms the combination of low cost access to skills and genuine 24-hour working. Since their inception in the late 1980s, call centres have become an established part of the UK economy. By 2007, the number of outsourced call centre advisers will have doubled, as companies look to outsource non-core competencies (Datamonitor 2003a). The issues researched in UK call centres, such as attrition rates, recruitment and retention, pay and conditions, skills, training and employee well-being, have become global issues, as has the gendered nature of call centre employment, the core issue of this chapter.

Although the emergence of call centres has stimulated research from a number of academic perspectives, whether or not women have access to careers in call centres remains a largely under-researched area. As a result, this chapter places the female call centre worker centre stage by asking, do call centres offer women an opportunity to progress into management positions? If so, what types of management positions are they, for example, do these involve an extension of the use of their 'soft' skills or something different? The chapter uses case study research to answer these questions. As call centres are a relatively new work arena, overwhelmingly populated by women, drawing extensively upon their 'soft' skills and increased flexibility, they are an ideal place for the exploration of the gendered social relationships between men and women in the workplace. Although call centres are overwhelmingly populated by women, the senior management team is invariably predominantly male.

The UK banking industry and call centres

The UK banking market reached a value of US$6,106 billion in 2002, having grown with a compound annual growth rate of 13.4 per cent in the 1998–2002 period. The highly competitive UK banking market has become increasingly consolidated as acquisitions have become commonplace and increasingly necessary for companies to maintain their market share. The recent growth in the UK banking market has been stimulated by the introduction of new methods of banking (the telephone and retail banking) and new technologies (Datamonitor 2003d).

The UK banking industry has been pivotal in the development and expansion of call centre working. The phenomenal growth in call centres has been attributed to the banking industry's early adoption of this new channel for service delivery, which was implemented in parallel with the reduction in the size of branch networks. This shift in channel delivery was inspired by the banking sector's desire to reduce costs and offer a more time-flexible service to customers, in an increasingly competitive market (Storey *et al.* 1999). A recent Datamonitor survey found that, in European retail banking, call centres are second only to branches in terms of generating business, with retail banking adviser positions accounting for nearly half of all financial services call centre adviser positions. Consequently it appears that banks have pushed a substantial number of enquiries into call centres making the corresponding cuts in the branch networks (Datamonitor 2002). As such, call centres are a product of complex, competitively driven spatial and organisational change within corporate structures (Bristow *et al.* 2002) being sites of both cost efficiency and customer orientation (Korczynski 2002).

In 2003, there were 435,000 call centre adviser positions in the UK and this has been forecast to increase to 517,000 by 2008, an increase of 18.8 per cent. Currently the UK is the leader in the European call centre market with a share of 31.7 per cent. Financial services is the leading sector in the UK call centre market with a share of 21.2 per cent in 2004, the manufacturing sector being the second largest and accounting for 11.4 per cent of market value. A large number of UK call centres are based in the South East of England and London, but many of these centres are smaller and more specialised than those in the North, Scotland, Wales and Northern Ireland (Datamonitor 2004).

The call centre market is technology-driven and new networking technology, including the Internet, has enabled companies to link together advisers in new ways to give them extra flexibility and lower overheads. Call centres have a gendered workforce, women comprising around 70 per cent of all call centre workers in the UK. They have a young age profile, with 63 per cent of employees aged under 30 years (IDS 2001). Despite the perceived risks (see Taylor and Bain this volume), offshore call centre positions are growing and by 2007 the number of outsourced call centre advisers in Europe, the Middle East and Africa will have doubled, as companies look to outsource non-core competencies. However, cracks are beginning to show in the outsourcing market over the tightening supply of skilled IT workers and the high rate of attrition amongst workers in the sector (Datamonitor 2003a, 2003b).

Gender stereotyping in call centres: women offer core skills

Frenkel *et al.* (1999) postulate that call centre work involves the use of different levels of contextual knowledge by 'frontline' workers, whose work is *different* from production and back office work because it is people orientated, is rarely completely routinised and is especially sensitive to changes in internal and external organisational environments. It is further argued that call centres demand key skills from their employees (Belt 2003; Belt and Richardson 2000; Callaghan and Thompson 2002; Korczynski 2001; Thompson *et al.* 2004). Consequently, the process of selection, recruitment, induction and training (to identify and shape social competencies) has moved centre stage (Belt 2003, 2004; Thompson *et al.* 2004) to pinpoint these communication and customer service skills and personality requirements (attitude and behaviour) that management see as critical differentiators.

The role of the call centre worker is both complex and simple, though worker profiles are skewed largely towards the lower rather than the higher end of the skills spectrum (Kinnie *et al.* 2000; Baldry 2002). The range of skills required include: handling customers professionally and efficiently under strict time pressures, whilst exercising discretion; conveying the right image, by being the 'public' face and first point of contact for the organisation; the performance of emotional labour (see Hochschild 1983); good communication and telephone skills and the ability to build rapport with the customer (friendliness, patience and active listening). Simultaneously, advisers are expected to operate information technologies (up to eight software packages using a combination of computer, telephone and Internet), navigate several knowledge systems and demonstrate a high level of technical understanding of the constantly changing product range. High standards of keyboard skills (speed and accuracy), numeracy, literacy, the use of appropriate grammar, spelling and product knowledge are also vital for this role where, increasingly, these skills are attributed to women, with a specific bias towards 'social' rather than 'technical' skills and competences (Durbin 2004).

The pressure to improve productivity and a broadening of tasks is forcing call centre managers to place an even greater emphasis on training. The most common type of training in Europe, accounting for 43 per cent of total training time, is in 'soft' skills, such as effective selling or calming angry clients. Call centres are now providing more advanced training to employees in the early stages of employment and using this as another filter for retaining advisers (Datamonitor 2003c). In the face of the consequences of increased competition, banks will migrate simple enquiries into cheaper, impersonal channels, like the Internet, leaving advisers to add greater value by dealing with the more complex enquiries and services (Datamonitor 2002).

Customers require call centre workers to demonstrate genuine commitment to giving a good service and to convey a feeling that they are being treated as individuals rather than the next customer in line (Korczynski 2001). Frontline service work can be distinguished from production-type assembly line work because of the uniqueness of the relationship between the customer and the front-line service provider (Batt 2000; Durbin 2004; Frenkel *et al.* 1999).

Call centre advisers and team managers sit in between what Frenkel *et al.* (1998) have described as users of 'lower-order' contextual knowledge (company-specific products and procedures) and higher-order contextual knowledge (conceptual understanding of different products, the market and the industry generally). Advisers are dependent on lower-order contextual knowledge for the routine aspects of their role but display a degree of creativity and social and organisational skills in their execution of more complex tasks.

Women's careers and call centres

Call centre careers have remained an under-researched area, with one notable exception (Belt 2002, 2004) where it has been argued that call centre employers are capitalising on gender divisions by actively recruiting women who they perceive will not be interested in promotion. As the call centre concept migrates from being merely a tool for simple, mundane transactions, to one that has the potential to sit at the heart of customer relationship management for many businesses, the role that women play in this process is worthy of investigation.

The call centre organisational structure resembles what Castells (2000) has termed a 'new organisational logic', the horizontal corporation that has moved away from its industrial organisational form (e.g. vertical integration) to enable it to adapt to the new conditions under 'flatter' hierarchies. Such structures have implications for career development and progression where the meaning of a 'career' has changed over time. Although careers can attribute coherence, continuity and social meaning to our lives (Colin and Young 2000), Storey (2000) depicts careers as being transformed into 'fracture lines' that have enabled the consideration of the longevity of multiple meanings of career. Littleton *et al.* (2000) describe the shift from 'bounded' careers – prescribed by relatively stable organisational and occupational structures – to 'boundaryless' careers – where uncertainty and flexibility are evident. A decline in evident career trajectories has been attributed to the impact of new technologies, the catalyst for technological displacement of employees, the restructuring of organisations, the increasing customisation of products and services and the impact of globalisation on local knowledge (Flores and Gray 2000).

There is a range of literature that draws out the ways in which 'career' is gendered and how women's careers are 'different' to those of men. Halford *et al.* (1997) demonstrate how gendered career paths have changed according to organisational restructuring and how men's career and employment conditions are promoted and prioritised over women's under these new conditions. Stanworth (2000) argues that although there may be some areas of convergence between the sexes in their experience of future work, men may continue to defend areas of competence and to dominate the high status and powerful occupational positions of the future.

Major sex inequalities have been found to persist at senior management levels in terms of the salaries and benefits offered to male and female staff and in access to certain favoured occupations. Male members of managerial elites hold firm views about the characteristics of the 'ideal' worker which inform organisational ideologies, such as practices concerning recruitment and promotion, thus creating an 'in' and 'out' group (Ozbilgin and Woodward 2004). Women's low representation in the management ranks has been explained through inadequate career opportunities, gender-based stereotypes and the 'old boy's network' (Oakley 2000).

Women's careers have been characterised by limited opportunities, unhelpful assumptions about commitment and capability, low paid part-time work, breaks of different lengths for child care and other domestic responsibilities (Wilson 1998). The new feminised career takes a number of forms, but incrementalism and credit accumulation are widespread strategies that, for some, represent a distinct resistance to masculine norms of long-term goal setting which subordinate the rest of life (Pascall *et al.* 2000). Female bank managers who had 'succeeded' in the rapidly changing organisational climate of the 1990s had often done so by changing their personal lives, an action which is characterised by a less traditional division of domestic labour. Opportunities and constraints mean that women construct their work-life biographies which result in their employment behaviour (Crompton and Harris 1999).

Analyses of women's careers occur, therefore, through competing and overlapping perspectives which fall, broadly, into three categories: cultural (family and feminine ideologies and organisational cultures), structural (family structures and organisational processes) and action (women's choices and strategies) (Evetts 2000). The cultural dimension includes the culture of management, which is usually perceived as strongly masculine. Beliefs of what constitutes 'good' management are in conflict with women's roles and responsibilities. For example, 'good' management is not perceived as involving caring, relatedness or connectedness. The structural dimension is closely related to that of the cultural and includes the institutional and organisational forms of work organisation, for example, the divisions of labour and department systems as well as promotion ladders and career paths within work organisations and professions. These form the context in which women's career choices are made and which differentially affect the occupational destinations and career trajectories of women and men. Organisation structures consist of promotion ladders and hierarchies of work positions that determine women's paid work and career experiences.

In the call centre context, where careers have been researched, there have been three evidently competing perspectives. First, it is argued that call centre workers' employment is organisation-centred rather than occupational (Kinnie *et al.* 2000, p. 133). Second, that careers are sectoral (between call centres) with massive horizontal mobility from organisation to organisation (Bagnara 2000, p. 20). Finally, that the different sectoral, occupational and organisational bases of call centre work arguably afford a range of different images of career, as well as of employee experiences (Watson *et al.* 2000). This is because the concept of career does have meaning within call centres, albeit in a very different sense than that used in more stable and less constrained employment contexts.

Belt and Richardson (2000) and Belt (2002, 2004) add a gender dimension to this discussion revealing a 'mixed picture' for women in call centres, with their careers ending at the supervisory level due to a general lack of managerial roles available, insufficient management training and, for women with caring responsibilities, difficulties in reconciling the demands of a managerial career with family life. However, Belt (2002, 2004) acknowledges that call centres are 'not entirely careerless' as several women are 'making it' to team manager positions while a few achieve higher levels of management.

Methodology

This chapter is based upon fieldwork analysis of four call centres that form a part of two of the UK's largest financial services organisations. The fieldwork was carried out between 1999 and 2001 and involved the use of semi-structured interviews, questionnaires and non-participant observation. A total of 114 interviews were conducted with key call centre personnel, ranging from advisers to Board directors in the parent organisations. The analysis in this chapter draws upon 33 semi-structured interviews with female team managers and 27 semi-structured interviews with senior managers (all of whom were male except one female at Bankco) in four call centres (two at Bankco, two at Finco). Observations and informal interviews with advisers in their day-to-day interactions with customers are also drawn upon (Table 7.1).

Interviews were the primary research method used with personnel at all levels in the call centres, the type of interview depending upon the research questions and call centre population. Structured interviews were conducted with team managers to complete a questionnaire but also to allow the interviewer to learn more about their role and the operation of the call centre more generally. All interviews at Bankco were conducted in a private office and at Finco, team manager interviews were conducted at the team manager's workstation, senior manager interviews in private offices. The interviews were used to establish the core skills required for team manager employment, to ascertain why women sought employment in the call centres and their career progression to date. Semi-structured interviews were conducted with senior managers in the call centres and parent organisations, to explore their perceptions of why women sought employment in the call centres and to identify the routes that they believed existed for the career progression of female team managers. Non-participant observation was continuously carried out throughout the fieldwork to gather information about the organisation of the work process and to observe interactions between managers and the workforce, as well as advisers and customers.

Gender and skills in call centres

Adviser and team manager levels in call centres are predominantly populated by women, who possess the required 'soft' skills; but are women

Table 7.1 Overview of Bankco and Finco case studies

	Bankco	Finco
Type of call centre	Financial services	Financial services
Technology platform[a]	Third generation	Third generation
UK location	West and North	South
Total number of call centre employees	4,000	700
Team managers:		
Total population	60[b]	51
Interviews conducted	22	11
Gender composition	75% female	75% female
Employed part-time[c]	4%	22%
Aged under 35 years	75%	89%
Married/living with partner	83%	33%
Dependent children	54%	22%
'Higher' level qualification attained[d]	33%	33%
Senior manager interviews	14 (100% male)	13 (100% male)

Notes
a Calls are handled in a seamless fashion across multiple physical points within one or two multiple call centres. The information technology will instantly recognise products in which the caller is interested and route the customer to the appropriate adviser.
b Team manager population 169 overall, but permitted to interview daytime staff only, of which there were 60.
c Within this full/part-time definition, there are just over 100 working parameters to cover the 24-hour, seven day a week operation: full-time (37 hours per week); reduced hours (25, 20, 16 hours per week); zero hours; annualised hours; fixed hour contracts; fixed rotating (alternate weekends); and flexible contracts (which must include 25 core and ten unsociable hours). Full-time working incorporates flexibility through the application of working parameters which operate on the principle of alternating start and finish times over a two-week cycle.
d Postgraduate/undergraduate degree; diploma; HNC/HND.

being held back because they possess these skills? This question resonates with Enloe's (1988) definition of 'patriarchal institutions' whereby call centres 'need' women because of specific skill requirements but like to keep them 'in their place'. When senior managers at Finco and Bankco were asked why they thought their call centres were overwhelmingly populated by women, a whole range of stereotypes began to emerge:

The 'return to work mum', employed on a part-time basis, was identified as the preferred archetypal call centre worker:

> I think females are far better at building empathy with people – without a doubt they are. I can range my best advisers on a number one, returning mothers working part-time, they just have that empathy, they're here for a fixed time and they do it and they probably enjoy coming in because they want to see and socialise with people as well. My poorer advisers are 18 to 24 year-old males.
>
> (Male Head of Call Centre, Finco)

I think that part-time is typically more attractive to women ... it's not a particularly well paid job and you know, it would be difficult to be a head of household and you know, be a primary earner and work in a call centre.

(Male Senior Manager, Parent Organisation, Finco)

Communication skills were also identified as important:

Communication skills, women are natural team leaders aren't they, they get on better with people than men do sometimes ... we want people who are able to cosset, care for our customers in a remote sense.

(Male Senior Manager, Parent Organisation, Bankco)

Um [long pause] I honestly don't know ... I really have no idea why it's predominantly female other than that the possibility that it is seen as a second income as opposed to a primary income so that, what, oh, I've thought of another reason actually ... the other reason is probably around the hours. Because there are a lot of reduced hours type roles, again it doesn't lend itself to a breadwinner situation. Now that sounds, that sounds as if it's sexist in that I'm assuming that the man is always the breadwinner in the family.

(Male Head of Call Centre, Bankco)

Male senior managers at Finco and Bankco characterise their ideal employees as women who are return-to-work mums who want to work part-time, who are able to empathise and communicate with the customer, to cosset and care for that customer and want to earn a 'second income'. Women are identified as 'attractive' options, as their employment needs allow managers more readily to plan to meet the peaks and troughs of customer demand.

In this sense senior managers are making assumptions about the role of women in call centres. It would appear that these call centres 'need' women because the public prefer a woman's voice, women fit in with the hours (due to their orientation to the family), do not expect to have a career and some are prepared to work for a lower wage. However, this stereotyping of women by their male managers is not only inaccurate but has also been found to impact upon women's career development (Wirth 2001). The crucial point is that call centres *need* women but also *use* women.

These stereotypes discussed previously can be challenged on a number of points, for example, team managers in all call centres were asked what attracted them to work in a call centre. The reasons cited are outlined in Table 7.2.

These findings indicate that initial employment in a call centre was not a 'career move' for almost half the respondents and that at least a quarter

Table 7.2 What attracted you to work in a call centre?

Reason	%
The opportunity arose	45
New challenge	24
Nature of the work	21
Hours of work	10

were looking for a new challenge. That women also sought call centre employment due to the nature of the work serves to reinforce the male senior managers' view of why women seek employment in call centres, even though this was not because of the hours, as most of the senior managers interviewed believed. Furthermore, advisers have an average target salary of £17,000.00 per annum, this rising to £25,000.00 and £18,000.00 per annum at Finco and Bankco, respectively, for team managers. These salaries hardly constitute a 'second income', especially when bonuses, incentives and financial benefits are considered. In addition, as 46 per cent of team managers at Bankco and 78 per cent of team managers at Finco have no children, this does not correspond with the 'return-to-work mum' scenario.

Although the data challenges the stereotypes of why women work in call centres, the reality remains that call centres need and use women because they serve the requirements of the patriarchal perspective and the institution in which they operate. Are these stereotypes serving to keep women out of management in call centres?

Career progression from 'frontline' adviser to team manager

Movement from adviser to team manager levels in the case study call centres is both a possibility and a reality for a large percentage of women. The extent to which women have moved beyond team manager positions, into middle and senior management roles in the call centres is assessed and the views of (male) senior managers are once again drawn upon to determine their perceptions of how women can move into more senior positions within the call centres and parent organisations more generally.

The team manager population has responsibility for the management of advisers. Key elements of the team managers' role are to motivate advisers, keep customer satisfaction levels high and serve the requirements of senior managers. This is achieved through the application of a mixture of 'hard' and 'soft' skills, such as maintaining performance levels and coaching and training advisers. Team managers possess the specific social and technology skills required for call centre employment, such as proficiency in the use of information technologies, teamwork, communication and learning skills. At both Bankco and Finco, the percentage of women who had progressed from adviser to team manager positions is high, at 79 per cent, indicating upward mobility at this level. Team managers at both Bankco and Finco are responsible for ten advisers each.

To map career progression from adviser to team manager levels, several areas of past experience and job mobility were explored. First, the percentage of team managers previously employed as advisers in the call centres was established to demonstrate the extent of upward mobility and the timeframe in which this was achieved. Second, sources of recruitment, previous employment in a call centre and job tenure were established. Third, team managers were asked whether they had experienced any barriers in their career progression to date. Finally, levels of career satisfaction were established.

Seventy-nine per cent of female team managers had previously been employed as advisers in their existing call centres, denoting an organisational rather than occupational career. In terms of length of service, 39 per cent had been employed as a team manager for less than one year, 30 per cent for one to two years and 30 per cent for between three and five years. Seventy-three per cent had been promoted from adviser to team manager positions within three years, 46 per cent within two years and 19 per cent within one year. Recruitment from within the call centre was the most likely route to team manager positions as 79 per cent of female team managers had taken this route. Twenty-eight per cent had previously worked in a call centre, predominantly at the adviser level (21%). This demonstrates that these team managers were not moving around the industry and, so far, had experienced promotion within their current call centres.

Team managers were asked whether they had experienced any barriers in their career progression and a range of perceived barriers emerged. Fifty-five per cent said that they had experienced barriers to career progression through a lack of management training, 30 per cent due to lack of career guidance and 21 per cent due to lack of promotion opportunities. The claim by Datamonitor (2003c) that call centres are providing more, advanced training is therefore questioned in connection with these team managers. Belt (2002, 2004) also found a general lack of appropriate management training opportunities for team leaders as well as a lack of management opportunities into which team leaders could progress. Lack of opportunities is, of course, linked to the flatter structures in call centres. This finding is not surprising as men have also been found to retain access to certain favoured occupations in other types of employment (Oakley 2000; Ozbilgin and Woodward 2004; Pascall *et al.* 2000; Wilson 1998).

Fifty-two per cent of team managers indicated that a lack of support from line managers also acted as a barrier to progress. Finally, 36 per cent stated that family commitments had acted as a barrier to career progression, a finding that parallels numerous other studies (for example, Belt 2002, 2004; Pascall *et al.* 2000; Wilson 1998).

Interestingly, when asked 'how satisfied are you with your career progression to date?' 52 per cent said that they were very satisfied while 48 per cent said fairly satisfied, denoting a fairly high satisfaction rating overall. Frenkel *et al.* (1998, 1999) also found high levels of job satisfaction amongst respondents in call centres which derived from helping customers and the social support that develops in the workplace. Over half were satisfied or very satisfied with the methods of control used and three-quarters said the controls helped them work better (Frenkel *et al.* 1998).

Overall, almost half the team managers interviewed did not see their initial employment in a call centre as a 'career move', even though just under half had been promoted within two years. A high percentage of female team managers had previously been employed as advisers, indicating opportunities for career progression, most notably after a period of three years. These team managers were very likely to have been previously employed as advisers and so would appear to take the more traditional career progression route from adviser to team manager. In terms of career satisfaction, a fairly high satisfaction rating is evident, even though many have experienced barriers to their career progression to date.

This analysis suggests that although women did not enter call centre employment for reasons of career advancement, there is evidence that this did occur within organisations as the vast majority of respondents were promoted within the call centre at a time when there was very little recruitment occurring externally (Kinnie *et al.* 2000). There also appeared to be very little sectoral movement between call centres (Bagnara 2000) as 72 per cent of respondents had not previously worked in a call centre.

There are evidently opportunities for women to move up to the first level of management, into team manager positions, where they continue to use their 'soft' skills. But where do they go from there? Examination of the call centre structures at Bankco and Finco reveals the extent of progression for women in relation to this study.

Team manager progression into middle and senior manager positions

Both organisations' call centres operate with flat structures, but with different spans of control at the middle management level. The span of control refers to the number of subordinates who report to a single manager and for whose work that person is responsible. In the organisation with a flat hierarchy, where there are many employees reporting to each supervisor, that person has a broad span of control. Although both Bankco and Finco have flat structures, the span of control is broader at Finco at middle manager level than it is at Bankco. Women are well represented at Bankco at the middle and senior managerial levels but under-represented at Finco.

Bankco has 34 *middle managers* in total, with a gender composition of 75 per cent female to 25 per cent male. Seventeen (50%) of these middle managers are ex-advisers and 19 (just over half) ex-team managers. This demonstrates that there is considerable scope for both advisers and team managers to move up into this position, which is a people management role with a span of control of 1:7. Finco has eight *middle managers* in total, with a gender composition of 25 per cent female to 75 per cent male. One female middle manager is an ex-team manager and at least three of the male middle managers were recruited from within the branch network. Middle managers at Finco have a span of control of 1:10, larger than that of Bankco, the role being a combination of operational and people management. There are fewer managers with more responsibility at Finco, and thus, the span of control is broader.

There are ten *senior managers* at Bankco, with a gender composition of 60 per cent female to 40 per cent male. One senior manager is an ex-adviser, five (50%) ex-team managers and six (60%) ex-middle managers. Again, this demonstrates the scope for movement up into this position, becoming more likely from team manager level upwards. This is an operational and people management role with a span of control of 1:4. Finco has two male *Senior Executive Managers*, this being a strategic role that involves working closely with the head of call centre and other senior members of the parent organisation. Each has a span of control of 1:4.

Bankco has two, male, *Heads of Call Centre*, each being responsible for several call centres that are geographically split but have a span of control of 1:5. The Head of the Call Centre is part of the decision-making group based in head office. Finco has one, male, *Head of Call Centre* who has a span of control of 1:2 and controls three prime operational areas – sales, service and Internet service – and is responsible for the development, implementation, monitoring and fine tuning of the call centre strategy.

Opportunities do exist for women to move up through the management hierarchy at Bankco. Women are well represented at middle management level where the need for 'softer' people management skills predominates. Women are also represented at the senior management level where operational capability and people management are combined. Women are, therefore, not positioned to make a major contribution to the strategic direction of the call centre. Team and middle manager positions are a springboard for progression to senior levels at Bankco. At Finco, the move up from team manager is much more difficult as there are very few places to progress into at the middle manager level (eight compared with 34 at Bankco). Therefore, the less graded the hierarchy, the less opportunities there are likely to be for upward mobility for women. As call centres are predominantly staffed by women, this has important implications for their managerial aspirations. Where women enjoy a critical presence, a mixed picture is revealed. Where roles tend to favour women, these are predominantly linked with people management. Where men and women are more evenly represented, women generally have a greater opportunity to influence decisions at the local rather than the senior management level.

Career opportunities for team managers: the senior managers' view

Senior managers at Bankco and Finco identified several routes into senior management roles. Senior managers at Bankco mentioned: movement into middle management level, where there is a 'reasonable turnover' of staff, thus creating opportunities for team managers; team manager opportunities in other parts of the call centre/support services; using recruitment into the call centre as a springboard into other management opportunities within the parent organisation. Senior managers at Bankco view the call centres as a 'fertile recruitment ground' for other parts of the business, the flat structure in the call centres being viewed as a barrier to progression within the call centre itself.

Chances of progression were also arising due to expansion:

> Well, more than 50 per cent of my managers started here on the telephones so one of the things I say to new advisers when they arrive and I do induction talks is to say that it is an environment where they can progress and it's about meritocracy, it's not about dead man's shoes and so yes, it is, the opportunities are there, we're expanding.
>
> (Head of Call Centre, Bankco)

Senior managers at Finco mentioned several routes into senior management positions: project work within the call centre (which will give access to Board and other senior members of the parent organisation); 'putting yourself out' by working long hours; 'putting yourself in difficult positions'; taking on more responsibility; having the 'right attitude'; exhibiting the right behaviours; a move into a managerial role at the same level in the parent organisation and subsequently into a senior management role; and taking external courses, such as an MBA or a marketing degree. Project work is one of two main routes for progression for team managers in the call centres. The benefits of this were made clear by the Head of Call Centre:

> You've got to get them out of the sort of bubble of the call centre and to start looking at other parts of the business where a project will give someone that push, they can get exposure to some Directors.

However, there are normally only three or four project teams operating at any one time, which can limit these opportunities.

The other route to progression is by 'putting yourself out' or 'having the right attitude', as explained by a director of the parent organisation:

> It's realistic and it's not. What are you doing in a call centre that actually says, I want to go further, are you taking, for example, an MBA,

are you doing a DMS, are you doing a marketing degree, are you doing anything externally? Now, if you've got the capability and all the other things and you actually put yourself out, work long hours, put yourself in the difficult situations, take on more responsibility, have the right attitude, demonstrate the right behaviours and do the external work that we've all done in going up higher then you will go higher.

These factors *could* have implications for women as taking external courses and working long hours is not always a possibility for those with extra caring responsibilities at home, such as children (54 per cent of female team managers at Bankco and 22 per cent at Finco have children). It would appear that there are a number of opportunities for promotion within the call centres, but with the 'flatter' management structure, these opportunities are limited. Progression via an organisational route outside of the call centre is probably therefore most likely at Finco.

Conclusions

This chapter has analysed the concept of career in the call centre context by asking whether this is gendered. It is evident from this study that call centres are not 'careerless' environments (Belt 2002, 2004) with no opportunities for career progression, as in both organisations, vertical and horizontal career pathways existed. Women are progressing into people management roles at the team manager level at Finco, but their progression stops there. This may be due to the broader span of control at the next level and a paucity of positions (eight in total) into which they can progress. Women have progressed up to senior management levels at Bankco and this could be due to the narrower span of control and more positions being available (34 in total) at middle manager level, which may act as a springboard into senior management (60 per cent of senior managers are ex-middle managers). However, these are predominantly people management roles, the decision making roles, once again, being filled by men. Flatter and broader organisational structures and broader spans of control are, therefore, preventing women from progressing into non-people management roles in the case study call centres.

A further structural reason for women's lack of progression is linked to a lack of women as role models at senior levels of the call centre and within the parent organisations' senior management teams. The importance of the women in these positions contributing to decision making processes and acting as role models for other women has been well documented (Davidson and Burke 2000; Kanter 1977; Rosener 1990, 1995).

From the cultural perspective (Evetts 2000) male senior managers hold stereotypical perceptions about the skills and ambitions of their female workforce and this is reinforced when women progress into people management roles. Traditional stereotypes and attitudes about women's abilities appear to limit their career prospects. Belt (2004) found that some team leaders felt they were perceived by their managers as either not interested in promotion or as unsuited to management roles and this also applied at Bankco and Finco. Belt (2004) also found that external candidates were recruited for management roles, but this appeared not to be the case here as Bankco's middle and senior managers were mostly recruited from within the call centres, whereas Finco's were recruited from within the branch network.

A range of perceived barriers to career progression emerged from the interviews with team managers. Fifty-five per cent said that they had experienced barriers to career progression through a lack of management training, 30 per cent due to a lack of career guidance and 21 per cent due to a lack of promotion opportunities. This questions Datamonitor's (2003c) claim that call centres are providing more and advanced training, while it parallels Belt's (2002, 2004) findings that a lack of appropriate management training opportunities for team leaders exists as well as a lack of management opportunities into which team leaders could progress. Men have been found to retain access to certain favoured occupations in other types of employment (Oakley 2000; Ozbilgin and Woodward 2004; Pascall *et al.* 2000; Wilson 1998), but these barriers remain relatively unexplored in the call centre context. Fifty-two per cent of team managers indicated that a lack of support to progress from line managers also acted as a barrier, but this finding has not emerged as an issue in the gender and careers literature generally. Accordingly it warrants further investigation. Finally, 36 per cent stated that family commitments had acted as a barrier to career progression, a finding that parallels numerous other studies, including those of Belt (2002, 2004) Pascall *et al.* (2000) and Wilson (1998).

In summary this chapter has investigated women's career progression in call centres from the structural and cultural perspectives. It has drawn upon the 'maleness' of management by exposing male manager's stereotypical perceptions of why women work in call centres. Women's low representation in management has been explained through the persistence of gender-based stereotypes where male members of the managerial elite group hold firm views about the characteristics of the 'ideal' worker which informs organisational ideologies, such as practices concerning recruitment and promotion (Oakley 2000; Ozbilgin and Woodward 2004). This warrants further investigation in the call centre context, especially as the male stereotypes presented have been challenged by the realities of why women seek call centre employment.

It has also demonstrated how barriers created by corporate practices, such as flat organisational structures, make it difficult for women to progress in call centres. Where women have progressed in the case study call centres, this is characterised by stable employment with one employer

(Kinnie *et al.* 2000; Ranson 2003). This is also evidence of gendered career paths based on organisational restructuring where, when women are promoted, their role remains predominantly focused on people management. Men's career and employment conditions are, therefore, promoted and prioritised over women's under these new conditions (Halford *et al.* 1997; Stanworth 2000).

Women are progressing into management positions in the call centre case studies at the first level of promotion from adviser to team manager. They continue to progress at Bankco, but into people management roles where their 'soft' skills remain a requirement. At Finco, the next step up from team manager is into a role that offers limited opportunities due to the small number of vacancies available and the broader span of control. The levels above these, have not, to date, been filled by women. This chapter therefore offers a mixed picture of women's progression in call centres, based on flatter organisational structures, a lack of female role models in management positions, the male stereotyping of women's skills and perceived barriers to career progression for women, such as training, promotion opportunities, lack of support and family commitments.

References

Bagnara, S. (2000) 'Towards Telework in Call Centres', www/euro-telework.org (accessed 2 June 2001).

Bain, P. and Taylor, P. (2000) 'Entrapped by the "Electronic Panopticon"? Worker Resistance in the Call Centre', *New Technology, Work and Employment*, 15(1): 2–17.

Baldry, C. (2002) 'Editorial: Work Technologies and the Future of Work', *New Technology, Work and Employment*, 17(3): 152–153.

Batt, R. (2000) 'Stratetic Segmentation in Front-line Services: Matching Customers, Employees and Human Resource Systems', *The International Journal of Human Resource Management*, 11(3): 540–561.

Belt, V. (2002) 'A Female Ghetto? Women's Careers in Call Centres', *Human Resource Management Journal*, 12(4): 51–60.

Belt, V. (2003) 'Work, Employment and Skill in the New Economy: Training for Call Centre Work in the North East of England', paper presented to the *21st Annual International Labour Process Conference*, University of the West of England, Bristol, UK, 14–16 April.

Belt, V. (2004) 'A Female Ghetto? Women's Careers in Telephone Call Centres', in S. Deery and N. Kinnie (eds) *Call Centres and Human Resource Management: a Cross-national Perspective*, Hants: Palgrave Macmillan.

Belt, V. and Richardson, R. (2000) 'Women's Work in the Information Economy: the Case of Telephone Call Centres', Centre for Social and Policy Research, University of Teeside, *Occasional Paper* No. 1: 7–22.

Bristow, G., Gripaios, P., Keast, S. and Munday, M. (2002) 'Call Centre Growth and the Distribution of Financial Services Activity in the UK', *The Service Industries Journal*, 22(3): 117–134.

Callaghan, G. and Thompson, P. (2002) 'We Recruit Attitude: the Selection and

Shaping of Routine Call Centre Labour', *Journal of Management Studies*, 39(2): 233–254.

Castells, M. (2000) *The Information Age: Economy, Society and Culture; Volume I: The Rise of the Network Society* (2nd edn), Oxford: Blackwell.

Colin, A. and Young, R. A. (2000) (eds) *The Future of Career*, Cambridge: Cambridge University Press.

Crompton, R. and Harris, F. (1999) 'Attitudes, Women's Employment and the Changing Domestic Division of Labour: a Cross-national Analysis', in R. Crompton (ed.) *Restructuring Gender Relations in Employment: the Decline of the Male Breadwinner*, Oxford: Oxford University Press.

Datamonitor (2002) 'How Can Banks Centre on Productivity?', http://dbic.datamonitor.com/ (accessed 2 December 2004).

Datamonitor (2003a) 'Offshore Call Centre Positions Growing, Despite Perceived Risks', http://dbic.datamonitor.com/ (accessed 2 December 2004).

Datamonitor (2003b) 'Indian Call Centres: Cracks Begin to Show', http://dbic.datamonitor.com/ (accessed 2 December 2004).

Datamonitor (2003c) 'Call Centres: Focus on Soft Skills', http://dbic.datamonitor.com/ (accessed 2 December 2004).

Datamonitor (2003d) 'Banking in the United Kindgom: Industry Profile.'

Datamonitor (2004) 'Call Centres in the United Kingdom: Industry Profile.'

Davidson, M. J. and Burke, R. J. (2000) *Women in Management: Current Research Issues*, London: Sage.

Deery, S. and Kinnie, N. (2004) *Call Centres and Human Resource Management: a Cross-national Perspective*, Hants: Palgrave Macmillan.

Durbin, S. (2004) 'Is the Knowledge Economy Gendered? Call Centres as a Case Study', unpublished PhD thesis, University of Leeds, Department of Sociology and Social Policy.

Enloe, C. (1988) *Does Khaki Become You? The Militarisation of Women's Lives*, London: Pandora.

Evetts, J. (2000) 'Analysing Change in Women's Careers: Culture, Structure and Action Dimensions', *Gender, Work and Organisations*, 7(1): 57–67.

Fernie, S. and Metcalfe, D. (1998) *(Not) Hanging on the Telephone: Payment Systems in the New Sweat Shops*, Centre for Economic Performance, London: London School of Economics.

Flores, F. and Gray, J. (2000) *Entrepreneurship and the Wired Life: Work in the Wake of Careers*, London: Demos.

Frenkel, S., Tam, M., Korczynski, M. and Shire, K. (1998) 'Beyond Bureaucracy? Work Organisation in Call Centres', *The International Journal of Human Resource Management*, 9(6): 957–979.

Frenkel, S. J., Korczynski, M., Shire, K. M. and Tam, M. (1999) *On the Front Line: Organisation of Work in the Information Economy*, USA: Cornell University Press.

Halford, S., Savage, M. and Witz, A. (1997) *Gender, Careers and Organisations*, London: Macmillan.

Hochschild, A. R. (1983) *The Managed Heart: Commercialisation of Human Feelings*, USA: University of California Press.

IDS [Incomes Data Services Ltd] (2001) *Pay and Conditions in Call Centres*, London.

Kanter, R. M. (1977) *Men and Women of the Corporation*, USA: Basic Books.

Kinnie, N., Purcell, J. and Hutchinson, S. (2000) 'Managing the Employment Relationship in Telephone Call Centres', in K. Purcell (ed.) *Changing Boundaries in Employment*, Bristol: Bristol Academic Press.

Korczynski, M. (2001) 'The Contradictions of Service Work: Call Centre as Customer-oriented Bureaucracy', in A. Sturdy, I. Grugulis and H. Willmott (eds) *Customer Service: Empowerment and Entrapment*, Hants: Palgrave.

Korczynski, M. (2002) *Human Resource Management in Service Work*, New York: Palgrave.

Littleton, S. M., Arthur, M. B. and Rousseau, D. M. (2000) 'The Future of Boundaryless Careers', in A. Collin and R. A. Young (eds) *The Future of Career*, Cambridge: Cambridge University Press.

Oakley, J. G. (2000) 'Gender-based Barriers to Senior Management Positions: Understanding the Scarcity of Female CEOs', *Journal of Business Ethics*, 27: 321–334.

Ozbilgin, M. F. and Woodward, D. (2004) '"Belonging and Otherness": Sex Equality in Banking in Turkey and Britain', *Gender, Work and Organisation*, 11(6): 668–688.

Pascall, G., Parker, S. and Evetts, J. (2000) 'Women in Banking Careers – a Science of Muddling Through?', *Journal of Gender Studies*, 9(1): 63–73.

Ranson, G. (2003) 'Beyond "Gender Differences": A Canadian Study of Women's and Men's Careers in Engineering', *Gender, Work and Organisation*, 10(1): 22–41.

Rosener, J. B. (1990) 'Ways Women Lead', *Harvard Business Review*, 63: 119–125.

Rosener, J. B. (1995) *America's Competitive Secret: Utilising Women as a Management Strategy*, New York: Oxford University Press.

Stanworth, C. (2000) 'Women and Work in the Information Age', *Gender, Work and Organisation*, 7(1): 20–32.

Storey, J. A. (2000) 'Fracture Lines in the Career Environment', in A. Colin and R. A. Young (eds) *The Future of Career*, Cambridge: Cambridge University Press.

Storey, J., Wilkinson, A., Cressey, P. and Morris, T. (1999) 'Employment Relations in UK Banking', in M. Regini, J. Kitay and M. Baethge (eds) *From Tellers to Sellers: Changing Employment Relations in Banks*, Cambridge, Mass., MIT Press.

Thompson, P., Callaghan, G. and van den Broek, D. (2004) 'Keeping Up Appearances: Recruitment, Skills and Normative Control in Call Centres', in A. Watson, D. Bunzel, C. Lockyer and D. Scholarios (2000) *Changing Constructions of Career, Commitment and Identity: the Call Centre Experience*, paper presented to the *15th annual Employment Research Unit Conference, 'Work Futures'*, Cardiff University, September 2000.

Watson, A., Bunzel, D., Lockyer, C. and Scholarios, D. (2000) 'Changing Constructions of Career, Commitment and Identity: the Call Centre Experience', paper presented to the *15th annual Employment Research Unit Conference, 'Work Futures'*, Cardiff University, September 2000.

Wilson, E. M. (1998) 'Gendered Career Paths', *Personnel Review*, 27(5): 396–411.

Wirth, L. (2001) *Breaking Through the Glass Ceiling: Women in Management*, Geneva: International Labour Office.

Community unionism in a regional call centre

The organiser's perspective

Al Rainnie and Gail Drummond

Introduction

In 2003 it was estimated that there were around 4,000 call centres in Australia (compared with 550 in 1998), employing around 160,000 staff. Potentially call centres are now relevant for all industries and all enterprises. In Australia location is highly concentrated. The majority are situated in Sydney and Melbourne where they represent approximately 70 per cent of the total multinational call centre market in the Asia Pacific region (Budde 2004). Nevertheless, Australia has been witnessing a tendency to relocate call centres from central business districts (CBDs). This trend has already been identified in the northern hemisphere, although infrastructure and staffing problems have limited the potential for cost advantages. Around 65 per cent of call centre costs are labour related, so where cost cutting is required, relocation and outsourcing go some way to accounting for the rationale for the much (over-) publicised offshoring of call centre activity (Budde 2004). The fear (or threat) that most call centres will relocate to offshore locations seems overplayed (see Taylor and Bain, Srivastava and Theodore this volume). The US research firm Datamonitor, in a report published in 2004, suggested that call centre employment in Australia would continue to grow at just under 10 per cent per annum until 2008. Despite the fact that India is expected to overtake Australia as the largest market for call centres in the Asia Pacific by 2008, factors including variable infrastructure reliability will militate against wholesale migration. The fact that over 50 per cent of Australian call centres offer languages other than English and have the capacity to carry out multi-lingual functions is also of increasing importance in location decision making (Invest Victoria 2004).

This chapter examines the experience of one union's attempt to organise an outsourced Greenfield call centre site in an old industrial region of South East Victoria in Australia – the Latrobe Valley. The area, about 150 kilometres South East of Melbourne, covers four small-to-medium sized towns and has a population of around 70,000. It supplies approximately 85 per cent of Victoria's energy needs through power generation based on open cast brown coal mining. The State Electricity Commission (SEC) was set up in the 1920s to provide for Victoria's power needs, relieving reliance on unionised black coal mines in New South Wales. During the post-war period the area attracted a reputation (somewhat misplaced) as a hotbed of union militancy and as a soft option for workers (SEC was said to stand for Slow, Easy and Comfortable). However, corporatisation, privatisation and restructuring in the mid-1990s had a devastating effect on both union organisation and the region itself. Unemployment rose to well above the national average and average wage levels fell below national figures. Social problems abounded and the area attracted a new and apparently equally unattractive reputation for social deprivation. Currently there are, however, some fledgling signs of renewal of both business confidence and employment within the region as well as in the local labour movement. This has been sparked to some degree by the arrival and subsequent union organisation of the call centre (Rainnie and Paulet 2005).

With particular reference to union organisation in call centres, this chapter picks up on Kelly's observation that we do not know much about how union organisation originates (Kelly 1998). In particular, this chapter can be seen as complementary to the work of Taylor and Bain (2003, p. 170) who outline the process whereby, in a UK call centre, from a collective interest among employees there arose the desire for union representation. Taylor and Bain (2003) argue that discontent is necessary, but not sufficient, for collective identification. The role of individuals (officials, activists, delegates, shop stewards and their political orientation) is vital in channelling the emergence of interest identification and a sense of injustice. This chapter focuses on the experience of the organiser who led a successful campaign to unionise the call centre. One of the strengths of the contribution is that it represents the organiser's story as a participant in the process through action research. Following Kelly's use of mobilisation theory (Kelly 1998, see also Danford *et al.* 2003, p. 17) it is argued that mobilisation depends on the ability and willingness of leaders to direct workers' sense of injustice toward management, and further, that such action often hangs on critical incidents which provide the opportunity for anti-management leadership. In effect, such incidents allow for employer-provoked collectivisation of employee discontent. As outlined later in this chapter, the Latrobe experience follows this pattern fairly closely. However, where Taylor and Bain (2003) told the story of the process of unionisation from the employees' point of view, here the authors tell a similar story, but from the point of view of the trade union organiser. Hence, this chapter fills an information gap, utilises the observations of the participants and complements the existing literature on trade union development in call centres, many of which are located on greenfield sites (Richardson and Belt 2001).

Call centres and local economic development

Richardson *et al.* (2000), in an examination of the situation in North East England, point to a number of concerns regarding the impact of call centres on the local economy. Call centres are an effective manifestation of the increasingly capital-intensive industrialisation of service sector work. Such work is highly intensive and routine. Furthermore, electronic surveillance provides the opportunity for the detailed control and discipline of workers. Production is highly specialised and the division of labour produces only a limited range of occupations, which combined with flat organisational structures restricts opportunities for career progression. The relative mobility of call centres, combined with the thrust of technological displacement, means that their lifespan threatens to be short, with international relocation feasible (see Taylor and Bain, and Srivastava and Theodore this volume).

Drawing on previous research, Richardson and Belt (2001, p. 73) list a number of locational characteristics for call centres. These are:

- advanced telecommunications suitable for data and voice transmission, capable of hosting intelligent network services;
- a plentiful pool of (often female) labour, which is skilled enough to carry out tasks required; labour costs are a factor but may be traded off against necessary skills;
- timely availability of property, together with low occupancy costs; together with the need to allow for expansion, this favours out-of-town or edge-of-town locations;
- fiscal and grant incentives;
- helpful and supportive development agencies; and
- access to good local transport.

Governments and regional development agencies around the world are marketing themselves as call centre locations with these characteristics in mind (for Australian examples see Dean and Rainnie 2005). Then again, as Richardson and Belt (2001, p. 74) point out, in attempting to emphasise difference, the regions that compete to attract call centres selectively harness positive images and data to present a sales pitch which simply reflects their sameness with all other localities involved in the bidding game.

Additionally, call centres are increasingly abandoning metropolitan locations and seeking areas further down the urban hierarchy. This is partly due to government action and incentives as well as rising land and labour costs. Because regions that are disadvantaged by distance and/or perceived economic uncompetitiveness can be attractive to call centres, this provides the possibility of unlocking under-utilised labour markets. According to Budde (2004), regional call centres can be 10 to 15 per cent cheaper to run than Metropolitan-based centres, but in rural and regional Australia they have to overcome problems of poor technology infrastructure, insufficient numbers of qualified staff and unreliable electricity supply.

The Latrobe call centre

The Latrobe call centre that is the focus of this chapter is important for the locality in that, by employing over 400 people, it rapidly became one of the biggest local employers. It was also relatively large by call centre standards being nearly three times the Australian average seat size. The call centre was outsourced by a large Australian organisation to a major company that is internationally owned and specialises in call centre management. The outsourcing company had a rigid non-union policy whilst the Australian client company is unionised, even in its other call centres. Wage levels for customer service representatives in this outsourced call centre were below the national average for call centres.

In discussing the location of the new call centre investigated in this chapter, a senior manager from the parent company explained that the new call centre was part of a restructuring of its services, which included the consolidation of a number of smaller centres. Consequently, no major claims about job creation were being made, as the senior manager outlined, 'it was probably aimed at cost cutting, but then we had this regional view overlaid on it'. Some positions were targeted to be filled by staff who were redeployed from various restructured units elsewhere. However, it was explained that 'because of the unemployment situation, we hope to get a few years with a lower turnover'. Turnover in the parent company's other call centres was running as high as 25 per cent for permanent employees with much higher levels being recorded for casual staff.

The call centre lay at the 'quantity' end of Taylor *et al.*'s (2002) continuum but, in fact, had two sections within it. One of these was a highly routine messaging service and did not include sales. The other section answered customer requests and endeavoured to provide 'solutions' and included sales. Both sections were highly monitored and controlled. Employees were expected to make new or further sales. Frontline employees performed integrated telephone and computer work, in response to inbound customer calls. They worked 6–8-hour shifts, in irregular weekly patterns. Employees took incoming calls for service enquiries (such as billing questions, product information and contract options) and, in finding customer 'solutions', they were also expected to make sales (transfers to higher value or new contracts, and additional products). Employees were organised into teams on the floor and each team had a leader who was responsible for supporting and mentoring team members.

For the Latrobe Valley, the new call centre's importance was couched in the familiar terms of job creation bringing the new economy to an old region. Despite the aim of employing more than 500 people, the parent company played down the job creation role as the opening was accompanied by the rationalisation of a number of smaller regional centres, including one in the adjacent town. What was not discussed was the level of subsidy that came with the location, the relatively low wage levels in the locality compared with metropolitan sites or how these factors might combine to reduce relatively high levels of labour turnover experienced at other parent company call centre sites. It was at this intersection of corporate restructuring with the local labour market and the community history of active trade unionism (Rainnie and Paulet 2005) that the struggle to organise the centre was played out, but first it is necessary to look in a little more detail at the relationship between the union and the community.

Locality, community and union organisation

An important debate has emerged with regard to the analysis of labour process development regarding the importance of place, locality and region (Rainnie and Paulet 2005; Herod 2002; Ellem 2001; Wills and Simms 2004). It is worth noting the points made by Ray Hudson in *Producing Places* (2001), particularly as he has some important comments regarding the nature of restructuring in formerly mono-industrial areas such as the Latrobe Valley. Hudson states (2001, p. 255):

> Places are complex entities: they are ensembles of material objects, workers and firms, and systems of social relations embodying distinct cultures and multiple meanings, identities and practices. They offer a setting in which production can occur and a way of organising production systems and the circulation of capital. Capital needs workers to be in their places to provide wage labour. National states seek to keep both capital and labour satisfied. Capital, labour and states and places thus exist in complex relationships... The processes of production of places and of social space are thus contested as different social groups seek to shape the geographies and landscapes of capitalism to reflect and further their particular interests.

Two assumptions underlie Hudson's (2001) approach. First, both place and space are seen as constituted out of spatialised social relations. Socially produced space is a product of social relations stretched out over space and materialised in a number of forms. Second, places only exist in relation to particular criteria. Places are not simply bounded but 'an unbounded lattice of power articulations constructed through and around internal relations of power and inequity. It is a discontinuous lattice punctured by structured exclusions' (Hudson 2001, pp. 256–257).

Places may be to some degree unbounded but, as Hudson points out

(2001 p. 265), particularly for mono-industrial communities dependent on a single employer a strong sense of place and deep attachment to it was often driven by necessity. Thus, for people living in such areas, an image of a bounded area served an important purpose. Such an attachment did not emerge overnight – rather it emerged slowly and haltingly out of deep divisions within the population. People thrown together chose to create a working class in a particular place. However, this attachment leaves open the potential for tension between place and class that remains problematic. Commitment is, however, affected by ongoing social practices (both progressive and regressive) which construct, sustain and transform the context within which economic, social and political life is reproduced (Hudson 2001, p. 267).

Moreover, the vulnerability to disinvestment (all places are prone to this to some degree) that is particularly true of mono-industrial areas can throw the tensions outlined earlier into stark relief. Opposition can be class based along the lines of strikes, factory/workplace occupations and/or alternative plans. Equally it can take the form of a communitywide alliance, in which case some groups foster the construction of a shared place-based interest that becomes the basis for action. Places in either of these cases are not coherent integrated wholes, but the image of a place can be used rhetorically to mobilise action. Such class alliance-based forms of intervention are increasingly driven by the devolution to the locality or region of responsibility for determining their own future. In particular, this can occur in peripheral regions leading to a focus on small firm formation (clusters), tourism and the attraction of manufacturing branch plants of Taylorised back-office service activities (such as call centres). To promote such development, images of locality that emphasise partnership rather than conflict are central. The image of partnership at local level does, however, come with the problem of setting locality against locality in the 'dog-eat-dog' game of regional development.

The contested nature of community and the changing nature of regional development policy leads to an examination of current union responses to developments in regions such as the Latrobe Valley. Much of the current debate on union strategy in the face of globalisation, membership decline and labour market restructuring has polarised around the 'partnership' versus 'organising' modes of unionism (Stirling 2004). There is a small but growing debate in the UK and in Australia on community unionism which draws heavily on US and Canadian experience (Lipsig Mumme 2003). As the intervention in the Latrobe Valley call centre was self-defined as a form of community unionism and drew heavily on forms of community intervention, as well as relying on given characteristics of the locality, it is worth exploring this debate a little further.

Community unionism

For Mathers *et al.* (2004, p. 12) current patterns of neo-liberal restructuring open up civil society and presents unions with a new set of strategic choices, particularly given the crisis of what they define as social democratic forms of trade unionism:

> The challenge is thus to develop new methodologies that are able to grasp the reciprocal relationship between trade union power and the wider society following the crisis and decomposition of social democratic trade unionism.

One such strategy may be community unionism, although as some critics have pointed out (Stirling 2004) examples outside of the US are few and far between, tend to be isolated, short lived and extremely local. The same may be true of Australia where it is still the case that the organising model has been taken up by few unions, in a strategic sense, in terms of resource allocation.

Wills and Simms (2004) have argued that community unionism itself is not new and has taken at least three different forms over many years of collective organisation in the UK, culminating in what they describe as reciprocal community unionism in which unions work with their communities rather than on their behalf. In the current context, Stirling (2004) following Wills and Simm (2004) and Lipsig-Mumme (2003) develops a threefold categorisation of community unionism:

> Firstly, there is community as identity. In this sense the union is the community in that the trade unionists are members of a community dominated by a single employer such as a pit village ... The community, the employer and union are entwined in a reciprocal relationship. Community action is likely to be oppositional to the employer and defensive – such as in a wage dispute or closure – and derived from a shared identity and a sharedness in the outcome.
>
> (Lipsig-Mumme 2003, p. 4)

Stirling (2004) ascribes such an approach to old mono-industrial communities (such as the Latrobe Valley). However, as already discussed, community unionism could take a number of forms in these circumstances and this approach does not take into account changing historical circumstances, nor the contested nature of notions of community.

Stirling's second category sees the community as a resource.

> This describes a situation in which both the unions and the community can share common interests and utilise each other as resources. The relationship will have peaks and troughs and periods of dormancy but

be associated with the longer term development of reciprocal relation-ships that are not necessarily focused on an organising campaign in a particular workplace or in a particular occupation but in building mutually supportive strategies.

(2004, p. 5)

This may be focused on longer term strategies with regard to particular communities (e.g. ethnic minorities) that challenge established union ways of working.

Finally there is the notion of the community as an instrument.

This describes a situation in which there is no necessary 'organic' con-nection between the union and the community but both might utilise each other for instrumental reasons. Particular workers are targeted and they are often those at the margins of traditional trade unionism. House calls, public meetings and the local media become key strat-egies alongside the mobilisation as supporters of the community organisations that represent the targeted workers. On the other hand, the community might be seeking organisational help for a petition or for accessing people with power. In both cases, this is generally a short-term relationship with clearly defined outcomes that can be met and dissolve the relationship.

(Stirling 2004, p. 5)

Community unionism offers a threefold possible advantage:

Firstly it is part of a strategy that can increase union membership. Sec-ondly by working with community organisations unions can achieve credibility with 'outsider groups'. Thirdly, the community itself pro-vides a point of pressure on the company that is useful in developing corporate campaigning strategies.

(Stirling 2004, pp. 12–13)

However, Stirling goes on to argue that in all these cases there is an assumption that workplace union organisation is 'in place'. Community unionism, for Stirling (2004), is dependent on effective workplace organi-sation which it can support but not replace. Our argument will be that this is a false dichotomy, the relationship between workplace union organisa-tion and community should be dialectical not dichotomous. This point is demonstrated in the Latrobe Valley case study. We now turn to an experience of organising in the call centre, an organisation whose work-force was an archetype of the new flexible economy, being composed mostly of young women with no personal history of union membership.

Unionising the call centre

In this section the authors draw on the experience of the union organiser. All the quotes are drawn from a series of interviews carried out over a two-year period as the campaign developed. The union organiser had chosen the region specifically to test some of the ideas that were emerging from the fledgling ideas of community-based unionism. This is the organiser's story in her own words. It represents the researcher as an observer and participant and fits within the action research approach to understanding in that it 'involves the researcher working with members of an organisation over a matter which is of genuine concern to them' (Eden and Huxham 2002, p. 255). For ethical reasons individuals cannot be identified and the call centre at the centre of the research is given the pseudonym of 'Newco'. The data presented are drawn from transcribed interviews, field notes and the organiser's own record of the process.

Survey data from the region (Rainnie *et al.* 2004) would seem to suggest that working people in the region were concerned about the nature of work and employment in the Valley, particularly its apparent casualisation, and remained relatively well disposed towards trade unions. Nonetheless, the once powerful Gippsland Trades and Labour Council (GLTC) was almost on the verge of extinction at this time.

Earlier organising attempts by another union through the activity of a locally based organiser had failed prior to the events outlined here. The organiser's initiative was triggered from within the call centre itself. An employee rang the union Membership Service Centre and asked for someone to come and talk to a group of five people about the benefits of joining the union. The meeting was to be held in the contactee's home. Through utilising friendships and community networks, membership began to grow. Contact with other union members in the area elicited more names and union membership grew again.

It is worth stressing again here Stirling's point (2004, p. 11) that there are not many living examples of community unionism upon which to draw, therefore, strategy and tactics emerged and evolved in practice. However, in any situation, organising is never going to work and form the basis for collective organisation and action without an issue that can be identified and attributed to the actions of management. In this case it was rosters and pay, rather than the much vaunted issue of surveillance which has greatly occupied the minds of call centre researchers to date:

> I went in over the pay slips that they actually weren't recording the actual hours worked. They were recording that they were working 38 hours a week. In actual fact they're required to come in 15 minutes earlier every day. So they're actually working more than 38 hours a week so their pay slips aren't true. So there was my breach. And it was hilarious. We went into the call centre. I think we signed up 30 people

in two days. People were just so keen to have a union inside that call centre.

Organisation and action spiralled and soon the organiser arranged the first of a series of significant barbecues, all held on a piece of land opposite the call centre.

> During July last year the few members that we had, along with some of our members from other work places in the Valley, arranged a BBQ in the local park. The members decided that the situation at Newco was not improving for them. Getting paid the correct amount each week seemed to be the biggest problem. Our members were ready to bring things to a head. Our members organised a stop work meeting for the following Monday morning. They used a phone tree to call their workmates at home and to let them know about the intended action.

> Forty-five workers marched up ... they've never done that, marched up and asked the HR manager for a meeting.

> She wouldn't come out and meet with them in the lounge together and it was absolutely stunning because one person said she bugged us every day when we were by ourselves and so you could see these 45 people feel the power ...

The action involved a large risk of going badly wrong as the organiser described:

> It was one of the scariest things that I had ever been involved with. I was worried that no one would walk off the job at the designated time. I was also worried that the employees could get the sack. Yet seeing the empowerment that it gave the workers was just amazing. This action had been a huge risk and luckily it paid off.

> They were all paid by twelve o'clock on Friday which is what we demanded.

The implications of the success of the action ran far beyond the call centre itself:

> Out of this action we got a lot of local support and media. The community was appalled that locals were not being paid correctly and getting the sack for taking too many toilet breaks.

This action was closely followed by the second barbecue, which was deliberately aimed at problems of the region as a whole and not just the call centre.

> I was still trying to increase visibility and community involvement. I got in touch with St. Vincent de Paul, a local charity. They do things pretty hard in the Valley because of the high level of unemployment. I arranged to hold a BBQ to help raise money for the soup kitchen that they run. Newco would not let us do this on site and we were forced to do it on the nature strip outside the call centre. The local council gave approval for the BBQ and also waived the fees that you would normally have to pay. Whilst speaking to the woman at the council, she arranged for me to meet her daughter, who worked at Newco. She signed up as a member. Our attempts to increase visibility were paying off, people were joining the union.

The thinking behind a community issue and community involvement went beyond a simple recruitment drive, rather it was driven by an understanding that with the increasing casualisation of the labour market (exemplified by the turnover rates in the call centre), the union would have to have relevance in all areas of people's lives if they were to retain an active and committed membership. Furthermore, in contrast to the old heavily male-dominated unions, whose image was not always positive even to those sympathetic to unions, integrating social justice issues into the activity of the union contrasted with the old view of such issues as optional extras to bread-and-butter workplace organisation.

The next stage in the recruitment drive was a weekend blitz. This involved the organiser and three graduates from the Organising Academy collecting names and addresses of people who worked in the call centre and then visiting as many as possible at home during the course of one weekend. The organiser was also aware that, as a tactic, a blitz was the flavour of the month but did not guarantee success:

> It's a bit of a silver bullet with Unions at the moment, the blitz idea. Everyone wants to try it for recruitment of course, because all the Unions are in trouble with memberships flagging, blah, blah, blah, so it's seen as a bit of a silver bullet, which is a shame ... Um, and if you're going to try a blitz, I guess you would try somewhere like Moe. That's a logical place to try ... well that's what I think anyway, because it's pretty open to Unions down here. It's not completely hostile, and if you're going to try, you have to try somewhere that's likely to succeed. And we thought that Moe would.

Almost inevitably, organising for the blitz required another barbecue, just over the road from the call centre.

We ... came up to Moe, had the barbie as you know, and invited a few people along to that. That was just to make it look legitimate, just to see if we had the support locally. The guys at the CFMEU who had done the Pilbara blitz – I wanted them to talk to people. Coz' we were all pretty scared, we were all pretty nervous. We're not trained sales people, so cold calling as it is, knocking on people's doors can be pretty scary ... we had the barbie, had way too many drinks, got up the next day ...

The involvement of local unionists from the regional Trades and Labour Council (GTLC) was important for two reasons; first, to draw on that experience; but second, because their presence at the barbecue signalled a local recognition of what the organising team was doing. The GTLC was an all but moribund organisation at this point and all attempts to get in touch with local union officials to try to gain support had, hitherto, failed or been ignored. But now organisers from a number of different unions came to lend support.

However, for the blitz to work, a degree of organisation within the call centre itself was necessary. A blitz, it was believed, would not work as the first stage of greenfield site organising strategy:

look that's biggest thing about a blitz, to have the visibility in the call centre before you do it. You have to have the visibility. I didn't want to get to people's doors and have to explain what the (union) was and what a union is, coz that's just going to take way too long, and we allocated 15 minutes for each person. Um we thought that was plenty of time really, and you've got to have the visibility and people have got to know who you are, and that was a huge help that I was so known in the call centre, like to look at 90 per cent of people know who I am.

The blitz in particular, and community-based unionism in general, are not substitutes for workplace union organisation, but can be effective in building and maintaining such organisation. Despite all their fears and apprehensions, the blitz worked very well, at least partly due to the traditions of the locality working in the unions favour:

Everyone was very friendly. I mean you'd get ... I'd knock on some doors, and the people would ask you in before you spoke! So you'd be inside and you'd be 'oh ...'; and they didn't work for Newco – I'd be at the wrong address! So it was very funny!! Or you'd knock on some doors and the parents would let you in and their kid wasn't home yet. And so they would ring their child, coz' they knew where they were and say 'Look, you've gotta come home – your Union's here.' 'Alright,' and they'd come home which was really cool, and they'd chase their kid, like if their kid was supposed to be at someone's house and they weren't there, they'd chase them to find them and send them home again! I think about 50 per cent of the people we saw were in their jammies, which was very cute and one of the interesting things was, you weren't just talking to the person that was at Newco – you were actually talking to the whole family. So you'd go into the lounge room and the whole family would be there, which was quite interesting too. Coz' a lot of the Dad's, if they were there, were saying how they were with the Union when they were at the power station, etc. So the family wasn't hindering, they were helping a lot. They were helping a lot – they're pretty incredible families down here.

By the time that the organiser left the region to take up a new post, the project looked like a success:

The project so far has paid off and is continuing to do so. We have eight workplace delegates in the Valley and two in Melbourne. We have a weekly WOC at one of the delegate's homes and regular phone hook ups with the Valley and Melbourne delegates. The delegates discuss what issues are happening in the two Victorian workplaces and campaign around them simultaneously. We have started lobbying the Latrobe Council for a council funded child care centre that could be placed in the community centre next door to the call centre and the list goes on.

Lots of hard work is still being done but it is getting to the point of, if you start work at Newco, you join the (union), this is the normal and O.K. thing to do. Like the members say 'it is like being in a special cool club'.

Furthermore, organising the Moe site had knock-on effects beyond the site itself and the Melbourne office:

I wanted to feedback to the rest of Newco across Australia and that's worked perfectly. Everybody in Newco in Australia knows about Moe and they know what goes on. Because as I said management's stupid! And will put emails to the whole company; 'what the (union's) saying is lies!' So they email it to the whole company which, of course, gives us visibility everywhere and gets us into the other sites. Moe has helped every Newco site in Australia. Every single site.

But the work is extremely labour and resource intensive and different from the work of standard union officials:

the work that I do here is different to what any organiser in the

UNION I think would do anywhere. Because I'm giving more of me, and more of my time, and you become more personally involved. You do, but I think you have to. To do proper community organising you have to. You do have to become involved with the community. And that involves giving more of yourself. But that's okay, it works.

However, the rewards can be great as far as the union is concerned and stretch beyond the call centre in question:

So for us working with the community has been a rewarding and challenging experience. The success of organising the call centre in the Valley has flowed through to other Newco workplaces in Melbourne, Sydney and Canberra. It has also helped with organising our more traditional areas of coverage in the Valley.

Conclusion

For Lopez (2004), community-based organising requires long term commitment of time and resources and a fundamental rethink of union organisation both locally and nationally.

In the case of the Latrobe Valley call centre, connection with the community paid off in terms of workplace union organisation. The two are interlinked and feed from each other, rather than being viewed as alternatives as Stirling (2004) seems to suggest. However this new form of organising demands a new approach from activists and organisers, in particular, and a real knowledge and connection with community life and issues beyond a narrow focus on workplace concerns:

Personally I feel that we have been accepted into the community. Going to the Valley for me is a very rewarding experience. I can sit in the call centre for four hours and have at least 30 people pop in to see us and just say hi. Even the security guard that Newco hire to watch us encourages workers to come in and see us. We are invited into member's houses for dinner and to see their kids perform at local concerts.

An understanding and involvement with the community outside of the workplace was crucial in determining the success or otherwise of the organising drive in the call centre. Such involvement is hardly new in Australia or outside of it. But the examples of union organising in and with communities, even within Victoria itself (for example the coal town of Wonthaggi in the 1930s) had largely been forgotten. In recent years Australian union organisation has tended to be highly centralised and institutionally focused. Even in areas, such as the Latrobe Valley, that did have a tradition of workplace-based union organisation, that organisation had often been conservative, male dominated and centred on a small number of heavy industries. The predominantly young female labour force in the call centre was employed by an American multinational company very different from the old SEC. Nevertheless, knowledge of, and involvement with, that union community allowed for a successful campaign amongst workers who are archetypes of the flexible and supposedly unorganisable workforce of the new economy.

In summary, there are positive implications for unions beyond simply the Latrobe Valley case. The movement of call centres from city centre locations to small town or offshore locations is being driven by labour cost concerns including issues of unionisation. Despite the widely perceived, but erroneous threat of all call centre jobs relocating to India, it has been suggested that in Victoria alone, call centre employment will increase by around 20,000 jobs by 2008 (Invest Victoria 2004). Hence, the Latrobe case would seem to suggest that the unionisation of this rapidly expanding non-traditional workforce is far from impossible.

On a wider front, the increasingly casualised and decentralised nature of Australian employment and industrial relations presents both a threat and an opportunity for union organisation. A focus on workplace-based organisation rather than state or federal level institutional structures requires a new way of organising. Fragmented, and small though they are, the examples of community-based union organising such as the Latrobe Valley case show that for unions, all is not lost. As Lopez (2004) points out, such an approach does not consist of a simple laundry list of tactics, rather it involves a process of change within the labour movement itself. However, there are signs that lessons to be learned from organising in sectors such as call centres could be applicable in the broader and equally rapidly restructuring areas of both the 'new' and the 'old' economies.

References

Budde, P. (2004) *Australia – Call Centres*, Paul Budde Communications Pty Ltd.

Danford, A., Richardson, M. and Upchurch, M. (2003) *New Unions, New Workplaces*, London: Routledge.

Dean, A. and Rainnie, A. (2005) 'Symbolic Analysts in the New Economy?', in A. Rainnie and M. Grobbelaar (eds) *New Regionalism in Australia?*, London: Ashgate.

Eden, C. and Huxham, C. (2002) 'Action Research', in D. Partington (ed.) *Essential Skills for Management*, London: Sage, pp. 254–272.

Ellem, B. (2001) 'Does Geography Matter?', paper presented to the Geography and Industrial Relations Forum, *15th AIRAANZ Conference*, Wollongong, January.

Herod, A. (2002) 'Towards a More Productive Engagement', *Labour and Industry*, 13(2): 5–18.

Hudson, R. (2001) *Producing Places*, London: The Guilford Press.

Invest Victoria (13.11.2004) News accessed at http://invest.vic.gov.au/News/News/CallCentres2004.htm (accessed 12 April 2005).

Kelly, J. (1998) *Rethinking Industrial Relations*, London: Routledge.

Lipsig Mumme, C. (2003) 'Forms of Solidarity', *Just Policy*, 30: 47–53.

Lopez, S. (2004) *Reorganising the Rust Belt*, Berkeley: University of California Press.

Mathers, A., Danford, A. and Upchurch, M. (2004) 'Opening Up Civil Society', paper presented to *Labour Movements in the 21st Century Conference*, University of Sheffield, November.

Rainnie, A. and Paulet, R. (2005) 'Image, Locality and Industrial Relations', *Australasian Journal of Regional Studies*, forthcoming.

Rainnie, A., D'Ubano, T., Barrett, R., Grobbelaar, M. and Paulet, R. (2004) *Industrial Relations in the Latrobe Valley: Perceptions and Reality*, Report commissioned by the Latrobe Valley Ministerial Taskforce Marketing Advisory Panel.

Richardson, R. and Belt, V. (2001) 'Saved By the Bell?', *Economic and Industrial Democracy*, 22(1): 67–98.

Richardson, R., Belt, V. and Marshall, N. (2000) 'Taking Calls to Newcastle', *Regional Studies*, 34(4): 357–376.

Stirling, J. (2004) 'Organising Communities and the Renewal of Trade Unions', paper presented at *Miners Strike 20 Years on Conference*, University of Northumbria at Newcastle, November.

Taylor, P., Mulvey, G., Hyman, J. and Bain, P. (2002) 'Work Organisation, Control and the Experience of Work in Call Centres', *Work, Employment and Society*, 16(1): 133–150.

Taylor, P. and Bain, P. (2003) 'Call Centre Organising in Adversity', in G. Gall (ed.) *Union Organising*, London: Routledge, pp. 153–172.

Wills, J. and Simms, M. (2004) 'Building Reciprocal Community Unionism in the UK', *Capital & Class*, 82: 59–84.

9   Agency and constraint

Call centre managers talk about their work[1]

Maeve Houlihan

Introduction

What do call centre managers and supervisors do? How are their roles
constructed? How do they talk about their work and what alternatives do
they see? This chapter explores the subjective realities of the frontline
management hierarchy in the call centre. Despite a growing literature theorising the work, work organisation and experiences of call centre agents,
research has to date said little about managers in this setting. This is an
important omission where exploration can offer some insight into the challenges and prospects for call centre work under consideration in this
volume.

This chapter will argue that call centre managers face a series of contradictory imperatives resulting from the collapse of service work into routinised production work and a heightened cost focus. As Taylor and Bain
(1999, p. 115) observe: 'In the drive to maximise profits and minimise
costs, call centre employers are under constant competitive pressure to
extract more value from their employees'. Further, call centre technologies mean that managers must motivate and manage performance within a
highly pressured and potentially monotonous work process (Taylor and
Bain 1999; Wallace *et al.* 2000). And yet, they are tasked to ensure the
quality of service provided to customers, and therefore, to offset cost and
quality imperatives – what Korczynski (2001) has described as the 'twin
logics' of achieving efficiency and providing effective customer service.
Within this struggle, the characteristic dual use of control- and commit-
ment-focused management strategies creates exacerbating tensions
(Kinnie *et al.* 2000; Houlihan 2002). So how do call centre managers navigate this difficult landscape?

This chapter aims to bring the subjective experiences of call centre
managers to the fore through the lens of two contrasting British call
centres, *Quotes Direct* and *Housing Helpline*. This discussion echoes
Frenkel *et al.*'s (1999, p. 14) depiction of the coexistence of two opposing
'images' of call centres – the regimented and the empowered organisation.
Explored together however, these cases highlight some of the common

tensions faced by managers in this sector, even across highly contrasting regimes.

Both cases form part of a wider ongoing study of the experience of working and managing in call centres. Data collection involved extended interviews and observation, and in the case of Quotes Direct, full personal participation as a call centre agent over the period 1997–1999. In both cases, contributors' accounts highlight the challenges and constraints that drove their daily practice, as well as the identity work in which they were engaged.

In taking this journey, this chapter illustrates two competing dynamics within call centre management. First, it highlights the variety, improvisation and innovations that (seek to) prevail in these managers' practices, but second, it suggests that the core architecture of call centre organising (task routinisation, automated workflow control and intensive performance monitoring) and its underlying tensions play a strong constraining role in call centre managers' autonomy to manage.

The manager in call centre research

Recent call centre research (Frenkel *et al.* 1998, 1999; Batt 1999; Korczynski 2001, 2002; Taylor *et al.* 2002; Kinnie *et al.* 2000; Callaghan and Thompson 2002) has played a valuable role in theorising work organisation and raised important questions about the nature of skills, and the personal consequences of routinisation, work intensity and technologies of control in call centres. Moreover, this research has been notable for its sensitivity to the voice and experience of the frontline worker. Yet within this corpus, managers' voices have become unusually silent. First, little has been said about managers' subjective experiences. Further, there is a tendency for this literature to present management as a homogeneous force, with implicit goals and outlooks centring on an implied 'relentless pursuit of control', ironically invoking the false belief that 'what management wants, management gets'. While call centre agents variously, comply, resist or negotiate their performance, managers, on the other hand, are presented as compliant and unproblematically performing their tasks (Willmott 1997).

But who is management? This study engages the perspectives of the frontline call centre management hierarchy, primarily the team leader, supervisor and centre manager. In many definitions, this grouping represents a 'middle management' structure, reflecting their accountability, but often, fundamental lack of decision-making power and autonomy. Call centres are centralised, typically 'flat' structures', wherein strategic decision-making generally takes place elsewhere while frontline management takes on a predominantly operational role. Nevertheless, from the perspective of the floor, this grouping *is* the management hierarchy and their practices and styles play a crucial role in how people see and experience the organisation.

Call centre managers 'walk the tightrope' between power and power-lessness (Kanter 1979), experiencing challenges to their upward strategic influence, yet pressure to deliver and implement policy downwards, manage the floor and control agent performance. At the same time, these frontline managers hold the strongest managerial connection to the customer interface. As key holders of the institutional knowledge base, they are expected to lobby for, support and protect the various interests of agent, customer and organisation.

Quotes Direct and Housing Helpline: a dialogue of opposites

'*Quotes Direct*' (QD), a 'commodity' insurance call centre based in the North of England, can be characterised as a chaotic and intensive production regime. Between 1997 and 1999, Quotes Direct employed approximately 600 staff. The call centre was divided into separate 'floors', respectively 'inbound' and outbound telemarketing, customer services, claims and a helpline. The participant study focused on the 'inbound' tele-marketing floor of this call centre, a group of 200 people whose job was to generate quotations and sell policies for home and motor insurance.

Work design on the inbound floor was governed by a complex array of rules and changeable targets. While utilising a high commitment-oriented language and a set of espoused values, its on-floor practice typified many widely echoed accounts of control-oriented call centre workflows (Taylor *et al.* 2002). Agent input was heavily constrained and scripts were strongly emphasised. The management structure at Quotes Direct took the classic form: teams of 12–20 agents were intensively micromanaged by team leaders, a room supervisor and a call centre manager who were, in turn, heavily restricted by centralised and unseen decision-making. This compounded the tendency for uncertainty and rapid changes in roles and operational guidelines for the call centre. The object of activity was thus interpreted locally as the achievement of service levels, sales levels and call length targets. In all these ways, Quotes Direct was typical of the many negative accounts of call centre organising and personal work experience.

In contrast however, '*Housing Helpline*' presents an uncharacteristic call centre story.[2] Housing Helpline formed part of a large third sector organisation, Housing Association, in the North West of England. Housing Association had been active for many decades, with over 100 staff and an elaborate divisional structure of satellite offices and central admin-istrative headquarters. Its activities included the management, develop-ment and maintenance of its buildings, tenants and business needs. It espoused a close relationship with tenants and rapid responsiveness to their issues.

In 1995, Housing Association began a two-year process of researching and developing a centralised call centre to handle communications with tenants, and to administer maintenance, rent payment and other needs. Under the guidance of Housing Association's IT director, a vision for this centre was developed which expressly sought to resist the 'typical' model of call centre organising. This aspiration was driven by the Housing Association's values, by concerns about service losses on the part of tenants, but also by the political reality that existing divisions and staff viewed the creation of a call centre as a threat.

The resulting call centre was located within Housing Association's headquarters, and went live in 1997. By 1999, it operated 25 seats with 40 staff, and ran 24 hours a day. Its presence within the heart of Housing Association operations meant constant interactions and physical move-ment occurred with employees in and out of each other's offices. Call centre staff shared canteen and front door facilities with the rest of the organisation, and several regular staff joined, or sought to relocate to the call centre, further increasing interconnectivity. The management team of the call centre comprised Housing Association's IT director who developed the centre, its strategic development manager and Housing Helpline's call centre manager. Two room supervisors and four team leaders, in turn, reported to the centre manager. While this accounts for a hierarchy, informal working relationships were very close, with ongoing daily interaction between all parties. The centre itself, partly due to its relatively small size, manifested a visibly open and familial atmosphere.

Underlying the two-year planning process was a concern to design supporting IT systems which both integrated the call centre with the wider organisation, and empowered call centre agents to work with dis-cretion on a wide variety of call-related tasks, using a shared and evolv-ing knowledge base. Thus Housing Helpline agents were actively constructed as problem solvers rather than information processors. In other respects, the call centre infrastructure was typical, as were its systems of performance monitoring and service level and call handling targets. The call centre manager, Barbara, was recruited from a senior post at a large commercial call centre and she shared the vision of an alternative call centre model. Due to its success both internally and externally, the Housing Helpline call centre had begun to market itself as a third party service provider to other organisations, and its business needs were expanding and changing rapidly.

An underlying theme of contextual constraints

The root conceptualisation of the call centre task – whether as production work or knowledge work – is the key to subsequent design and implemen-tation decisions. A limitation of the production/machine model is its neglect of the organic nature of human communication and interaction (Leidner 1993; Cameron 2000). A machine-like construction risks monot-ony, frustration and alienation, as Anna from Quotes Direct describes:

Being on the phone all day, you are blind, you are deaf, you're numb!
You don't know what time of day it is . . .. You're a vegetable . . .

(Anna, Call Agent – Quotes Direct)

Housing Helpline sought to pre-empt this dynamic by constructing the
job of call agent with skill and ethnomethodolgical competency in mind.
As the Call Centre Manager recounts:

We've tried to create a situation where people are enabled to solve
problems, to run with whatever it is the caller needs, and not just to
process information. [. . .] our staff have got a bit of autonomy in their
workload because they've got to find information for themselves,
whereas with the *dumb terminal*[3] you can only do what it is asking you
to do.

(Barbara, Call Centre Manager – Housing Helpline)

However, contextual dependence of Housing Helpline reveals the
fragile balance between 'regimented' and 'empowered' models of call
centre work (Frenkel *et al.* 1999, p. 14). Notwithstanding its greater
emphasis on service quality and agent autonomy, pressure from stakeholders, service level targets and unpredictable call volumes continually pressured Housing Helpline to routinise and tighten the reins of its
management strategy as echoed by an incident Simon described:

I was at one of our Board Meetings, and during the report on the call
centre one of the Directors joked 'lets call them now to test them!'
So they did. Right there on the speakerphone. And the call rang and
rang. And then it switched to the automatic 'please hold' message.
They got through in the end but that put the pressure on. I've committed to the call centre recovering its capital investment and we're
already exceeding our targets, but there is a constant need to prove
and justify our expenditure, our size, our budgets and our service
and that's made even sharper by [internal politics]. We've got to
provide a service that satisfies the internal and the external customer
and yes that's not best measured just in terms of volume but customers prefer to get an answer than to get a particularly good soft
skills answer, so that's why at the end of the day it has to be about
figures too.

(Simon, IT Director Housing Association)

Here we see a fundamental challenge to re-conceptualising values
about quality, service and roles. Balancing service level targets (implicitly
call answering times) against service quality (implicitly time spent with
each caller) is an intrinsic challenge to service provision due to unpredictable and erratic demand levels. Cost-driven pressures for lean staffing

leave the minimum of organisational slack, pressuring managers to reduce the time spent on each call by standardising, and moving back towards a machine model of production. As Simon goes on to say:

> I've seen enough of what goes on in other environments to make me careful about putting more pressure on staff in terms of productivity. It would be nice to have some spare capacity but there is this tension to maximise the benefits for the organisation in financial terms [...] And every year when we do the budget, I'm thinking this is quite tight ... how can I increase my staffing before they [the board] say hang on a minute?

Without strong managerial autonomy, and strong lines of support at higher management and board level for the espoused values around service, this organisation struggles to create the organisational space to realise and maintain its intentions of an alternative call centre model.

Stories such as this illustrate how the conflict between standardisation and flexibility, service and costs, and control and commitment strategies that lie at the heart of the call centre construct play themselves out, and suggest the inevitability of 'management by trade-off' despite the desire to do things differently. These embedded conflicts drive a tendency to inconsistent management strategies, so that for example when HR interventions such as teams, participation and TQM are introduced, their application tends to be uneven, superficial (Wallace *et al.* 2000; Houlihan 2002) temporary (Batt 1999) and conflicted (Knights and McCabe 1998; Kinnie *et al.* 2000). Thus, at Quotes Direct, the espoused HRM techniques comprised the valuing of employee involvement, yet lean staffing and call volume created an intensive and stressful environment and blocked the capacity for employee feedback. Similarly, at Housing Helpline, efforts to redesign work fundamentally came under constant downward pressure to normalise.

As Korczynski has stressed, the interaction between bureaucratic control and customer service objectives is a fragile balancing act – 'constantly switching emphasis between one and the other, trying to establish a form of equilibrium that they never quite reached' (2002, p. 92). These tensions underpin the work of call centre managers, forcing them to act as *translators* between the ideal and the real. They find themselves managing meaning and reframing inconsistencies as illustrated by the following quote from Barbara:

> Well actually we're having to pull it back a bit because we've gone too far with the flexibility and soft skills and their ordinary skills are dying – people aren't as good on the system as they are going to have to be. So we're pulling in the reins.
>
> (Barbara, Centre Manager – Housing Helpline)

In assessing this dynamic of switching back and forth, sometimes quality, sometimes quantity, Korczynski (2002) argues that these tensions cannot be reconciled, but must coexist. However the dynamics discussed in this chapter lead me to argue that the structural constraints and root assumptions surrounding the call centre model significantly predispose this efficiency-versus-service balancing act towards a cost and control character, rather than a harmonious state of coexistence.

What role does management play in these outcomes? The market-driven nature of the call centre sector leads to rapidly changing goals and processes and faces managers with considerable levels of uncertainty and contradiction. Understanding how they reconcile these conflicting messages and changing goals gives clues as to how they see the system and their own roles within it. What then do these managers have to say for themselves? These issues are explored in the following section.

The subjective manager

In this study the power attributed to call centre managers by the agents they supervise contrasted heavily with that reported by call centre managers:

> I wish we could have more power ... more ability to change things ... Power is the wrong word really. But ... if I had more influence on the people above ... Not just [the call centre manager] ... People that actually make the difference to the company.
>
> (Tina – Team Leader, Quotes Direct)

One of the significant findings from field data is the degree to which managers distanced themselves from influence and control in their accounts of themselves and their work. Rather, they pointed to the factors they saw as controlling their decisions and actions, for instance employee and customer behaviours, market demands, competition, the technology infrastructure, organisational codes, practices and technologies, and even other managers. As Hales (1999) suggests, far from the notion that 'the organisation is the manager's adventure playground', these managers strongly referenced the limitations and constraints they saw to be framing their work.

It is already clear that management at both centres faced ongoing contextual frustrations in relation to the core tasks of the job and the surrounding environment, particularly where lines of support were concerned. At Quotes Direct the formal system provided a demanding set of targets and role instructions. Managers saw their role as one of motivating and managing performance, yet the systems they worked within often antagonised this, forcing their emphasis towards control.

In the past it was more down to man-management, whereas now I feel as though the personal touch has gone away.

(Barbara, Centre Manager – Housing Helpline)

This was also observed by Jackie, a team leader at Quotes Direct:

Team leaders are the highest paid in the room, apart from the manager, and what do we do? We administer breaks. We've been trained, we can motivate, we can sell and we can use all the computer systems ... and they have us administering breaks. I've suggested we employ someone to do that, to let us get on with our jobs. The company won't have it though, and why won't they have it? Because there is no one to take it to that will listen.

In some respects, the technology architecture of the call centre predetermines management practice. Call monitoring and the routine collection of behaviour statistics suggest strong opportunities for micromanagement (for instance the notorious measurement of toilet breaks) that easily come to dominate. Although enmeshed in this system, managers are not unaware of its shortcomings. As Barbara comments:

I think that's overkill really, and I don't think it's necessary at the end of the day. They are human beings. As managers and team leaders we should be able to manage their down time to a satisfactory level that we don't need to be monitoring them that closely ...

(Barbara, Centre Manager – Housing Helpline)

Making choices

Sometimes frontline managers framed their role as being an intermediary between the letter of the law and what they felt should be its spirit. However, this often means disregarding formal job descriptions, using discretion and judgement, and risking censure. Thus, managers find themselves faced with a balancing act. They learn that achieving one target will lead to a penalty on another. They learn to recognise 'rules within rules', which reflect the contrast between formal requirements and the pragmatic negotiation of daily order. As a frontline manager at another research site commented:

If I achieved 100 per cent on our service levels, I'd be asked why. And next month, I'd be asked to cut staff numbers.

(John – Team Leader, Call Centre Bureau)

It appears that the job of the effective manager is to mediate and counterbalance the formal system, for instance drawing upon 'unofficial rewards' to make the system run smoothly – such as time away from the phones for 'creative' activities such as making team banners and room decorations, sending someone on an errand or finding more subtle means of maintaining employee and organisational dignity on a case-by-case basis. Yet such behaviour struggles to prevail in the face of heavily pre-scribed formal role expectations. Dalton (1959) observed that managers continually face 'moral burdens' in their struggle to reconcile 'recurring problems' and tensions. He described the personal, internal struggles of the manager and the struggle between 'the impersonal organisation and personalising individual' (Dalton 1959, p. 271). The frontline managers at Quotes Direct frequently talked about not being allowed to do the job they were employed to do, and they emphasised, the job they would *like* to do. In their terms, this means mentoring, communicating and developing people.

> No, the company doesn't want you to assume these roles. They want you to just do what they perceive is your job – to make their budgets and figures look better.
>
> (Jackie, Team Leader – Quotes Direct)

At Quotes Direct, the focus on managing the board and 'getting the figures right' prevailed to the point where taking agents off the phones to conduct even stipulated coaching sessions required, in one manager's words, 'sneaking around'. To undertake their jobs in this standardised and intensified environment, managers found themselves utilising creative workarounds and system 'blind-spots' (Callaghan and Thompson 2002), and colluding to manipulate an administrative system that, rather than being their servant, had become their master.

Tina recounts many examples of this, including secreting her team off the floor for a long overdue team meeting before they could be spotted and ordered back to the phones. Such collusion can be a pragmatic, even necessary strategy, though not always a comfortable one.

> My role is rewarding because you can work closely with staff and see the performance developing. That is really good. The thing that isn't rewarding is the fact that you've got to, you know, *twist* the system to get what you want out of it. And that's what is stressful ... it's not being able to do your work the way you should be.
>
> (Tina, Team Leader – Quotes Direct)

While the rational construction of the call centre system suggests deper-sonalisation, human personalities, power struggles and politics inevitably shape its enactment. This can also mean that good practice goes unnoticed and unshared. Such feelings of isolation are exacerbated by the tendency to use inter-team and inter-departmental competition to enhance performance, and managers regularly find themselves pitted against each other as a result.

Managing identity

It is worthwhile to examine the extent to which these managers chose to mediate all these pressures, the variety of strategies adopted and how these relate to their personal identities as managers. Here, Jackie accounts for her ambivalence regarding her role:

> Walking around with a clipboard doesn't do anything for me. I'd make more money and have less hassle if I just went back on the phones, but there is something in me that is always looking to the next level.
>
> (Jackie, Team Leader – Quotes Direct)

Jackie's approach to leading her team was one of practical mediation. For her, the business came first, and when new policies or practices seemed to clash with this, she challenged them on a number of occasions. Jackall (1988) reflects that the issue of identity is pivotal but fragile, in the management context. Inevitably the managerial role leads to clashes between self-image and some of the tasks managers are asked to do. These 'invitations to jeopardy' present the manager with a choice. Failing to compromise one's values in the political reality of the organisation can raise 'jeopardy' for the manager's career as judged by peers and superiors. Learning to choose which battles to fight, and how, was a process of learning to negotiate the politics of the system.

> Quite a lot of people here, you know, don't like me, but I don't care. Well I do care. But I find a lot of people confide in me. And other team leaders know that, and there's a bit of animosity against me for that. But then it's a case of working out what to do. If someone has a problem, going to the call centre manager doesn't get you anywhere. And if you go the HR Route, it's made formal. You have to get around everybody's little quirks, you know, which is awful. But it is better and quicker to do something informally, like go straight to the underwriter, or to a trainer, and then come back with it sorted and tell them here 'this is what I've done'. You do get into trouble for it, for going over heads. But at least you've caused some action.
>
> (Jackie, Team Leader – Quotes Direct)

At the frontline, there is keen awareness of the trade-off between system compliance and counterproductive outcomes, and the resulting effects on staff, customers and the business itself. It is at the interface this tension between the rules and the realities gets played out – that quality

and quantity, standardisation and flexibility, and control and commitment
– face each other.

Emerging from these managers' accounts is a sense that the stances
they took depended greatly on their own individual interpretive reper-
toire, reflecting the 'sort of people they are', but also their organisational
context and the sort of experiences and expectations they had of it. Managers who had themselves worked as agents tended to feel a keen understanding of the pressures absorbed by agents, which sometimes led them
to struggle due to the formal system within which they were required to
work.

> I feel the team ... need me [long pause]. If I was to throw the towel
> in, nothing would change for them, would it really? But I'm not
> going to let [the management system] win. That sounds like it is per-
> sonal, but yes. It can be a bit of a battle of wits sometimes. And on
> the other hand, I can remember being an agent, and I can remember
> needing help, and I can remember thinking 'I don't get paid for *this*
> stress ... why should anybody put you through this?' ... I feel as
> team leaders, we've got an obligation to make the staff feel comfort-
> able at work. They shouldn't be sat there worrying. And I think
> from a business point of view as well, this is where your hand's get
> tied. You want to see that point of view, particularly with the
> service levels, but then you see the other side of it which [the
> management system] has not seen, and that's where you get
> frustrated.
>
> (Terry, Team Leader – Quotes Direct)

At the end of the day, each manager remained accountable for
performance, and coping with this necessitates a certain distancing from
such sympathies. Focusing on the business was an effective way of reconciling conflicting emotions.

> Yeah ... there's stress. There's stress in being torn in so many differ-
> ent directions, ... trying to help other people that might be stuck,
> while doing the breaks, while being on the phones, while helping your
> own teams ... but you just do it. It's my loyalty to my employer, which
> is probably quite an old fashioned attitude I would think, but at the
> end of the day if people are not answering calls, and looking for busi-
> ness opportunities, and businesses don't exist anymore, there are no
> wages ... Alright, I might be a very small cog in a very, very big wheel,
> but I can say I've done my part, that keeps the whole thing going, you
> know, keeps the wheels of industry grinding ...
>
> (Helen, Team Leader – Quotes Direct)

Networks of support?

It is clear from the above that in a great many ways managers at Quotes Direct were forced to struggle to do the sort of job they believed in. Their effectiveness or ineffectiveness at balancing organisational tensions was compounded by the support systems available to them. Ironically, they found that the services and support they provided to their teams were not so forthcoming for themselves. The accounts of Quotes Direct managers speak of this neglect, in strong contrast with the accounts from Housing Helpline. Together they illustrate the difference support or its absence can make.

At Housing Helpline new team leaders shadowed existing ones for several weeks, gradually swapping roles from shadow to shadowed, until they were satisfied with their knowledge.

> Becoming a supervisor ... from managing the stats to knowing the answers to all the questions and being a coach, that was a big challenge. But working with [the call centre manager] is brilliant because she is so supportive. Her door is always open so I can go 'how would you go about this?' and she'll say 'well what have you done so far?' and then you start thinking of things together, and then you realise that's how to handle situations with the agents. She's coaching me the way she wants me to coach them. That's what they do here. We've all got the ability and we work hard, but that ability comes out because the back up is there for us.
>
> (Denise, Centre supervisor – Housing Helpline)

Yet in contrast, at Quotes Direct:

> When I started, I had half an hour at the Call Centre Manager's desk being thrown bits of paper. That was it. We're meant to have regular team leader meetings with each other, but haven't had one for about eight weeks. I have a meeting once a month with the Room Supervisor, a formal meeting to report back on all the statistics but I just ring her at home or she rings me if there is something we need to discuss.
>
> (Jackie, Team Leader – Quotes Direct)

Thus managerial support structures often went by the wayside at Quotes Direct, in contrast with the extent of codification and prescription of the job. Lack of formal training among call centre managers has been acknowledged by other researchers (Lankshear *et al.* 2001; Belt *et al.* 2002) and is symptomatic of rapid growth and short term thinking, as reliance on ad hoc and self-initiated skill acquisition risks narrow and fragile skill bases. This may contribute to the tendency to a glass ceiling at supervisory level. As Helen recounts:

I'm monitored purely on my team's performance. Now I've done detailed development plans for all my team and I can show you the evidence for everything they've done and all the progress they've made. Why can nobody do that for me? Who is developing me? To me something is missing – there's some sort of vital link in all of this, that's gone ... if that is available from me to my team – what is available from *my* management – for me?

(Helen, Team Leader – Quotes Direct)

Unsurprisingly, these team leaders reported both a lack of support, and a sense of frustrated isolation and often alienation.

I'd like to think what I do for my team is recognised, well I suppose it is in their stats, but you don't get a lot of support or feedback about how you are doing as a team manager – except when things go wrong or the stats are down, then they are all over us like a rash. But the rest of the time, you could be working your heart out, or just coasting, and I'm not sure they'd really care. I see a lot of my colleagues choosing that option. You know, just come in, do what they tell you to do, stay out of the way, and then leave.

(Terry, Team Leader – Quotes Direct)

This was a demanding context for these managers. It would be understandable for managers and supervisors at Quotes Direct to transfer this lack of support to their own teams and this was observed at times. However, more often, as these accounts suggest, the call centre managers tried to mediate the system and make things easier for their staff.

Interpreting outcomes

The stories these frontline call centre managers tell about their relationship with their work, and their observed practices, suggest a fundamental goodwill and flexible approach to negotiating daily order. However, while organisations rely on individual ability and willingness to cope, this risks personal disillusion and burnout, and the silencing of negative information. The role of frontline manager as buffer, facilitator, interpreter, mediator, advocate and support worker is relied on as an organisational coping strategy, but this reliance may quickly become self-perpetuating and regime sustaining.

The challenge for call centre managers remains to manage employee performance, but more informally, to manage the tensions and conflicts of the system in which it is embedded. This key role is played behind the scenes, often improvised, and certainly varied. The exploration of these accounts has conveyed that this work is poorly supported, suggesting the need for greater recognition of the effect and impact of frontline managers, and for greater support for and valuing of what they do. The story emerging is one of frontline managers who feel relatively little power to affect upward change or influence the system they manage. They cope by adapting at the edges, making small changes and personal concessions to make the system more bearable. Nonetheless, these frontline managers have a role, albeit one that they feel distanced from, in constructing the context within which agents work.

Moral mazes

The technology and performance management systems of the call centre perpetuate the illusion that managers have the tools they need, while obscuring 'ambiguities, uncertainties, contradictions and moral mazes' (Hales 1999; Jackall 1988). At the end of the day, managers are charged with responsibility and thus accountability for their actions, but more particularly the collective actions of their staff. As this accountability turns into 'liability' (Hales 1999), particularly in a defensive culture such as Quotes Direct, the manager is easily driven to conform to institutionalised routines and micromanagement. As Hales suggests, managers find themselves using the prevailing resources and cognitive rules about how call centre work should be organised and managed and consequently: '... by selecting, constituting and disseminating information through these rules and resources, managers weave the very webs of information in which they are entangled' (1999, p. 44).

The propensity for these managers to influence the system is shaped by their own attitudes, but moreover their context and the political realities of efficiency demands and pressure from stakeholders (Dalton 1959; Jackall 1988). In the Housing Helpline case, managers were conditionally supported by the wider organisation to do things differently. The evidence from the cases examined here is that the presence of strong guiding leadership at Housing Helpline, and its absence at Quotes Direct, framed the ways that these respective managers read the scope of their roles. Both, however, were conflicted. At Housing Helpline, for example, this insulating guiding vision had limits:

> Because at the end of the day, we came with job reductions. And the one thing they can see is that if the call centre can do what they were doing, and take their role, then what is left for them?
>
> (Barbara, Centre Manager – Housing Helpline)

At Quotes Direct, the absence of leadership created a vacuum that left managers without a sense of alternatives and strongly drove a tendency to defensiveness and isolation. However, it also freed up space for individual managers to negotiate their own interpretations. This was not coherent, and it struggled to thrive, but it did make a difference for some agents.

Accounting for difference

What accounts for the contrasts between the experiences of the two call centres explored in this chapter? Despite their differences in size, the call volume handled by agents at both Housing Helpline and Quotes Direct was similar, both ranging between 80 and 120 calls a day per agent. Yet their management styles, and certainly their work cultures, were dramatically different.

Earlier it was observed that a key foundation as to how a call centre operates relates to the ways in which the core activity is constructed. This, in turn, shapes decisions as to how to manage. At Quotes Direct, agents were constructed as production workers, processing scripts of which it seemed they needed to know nothing. This implied that they could be managed like *machines*. In contrast, Housing Helpline constructed their agents as *problem solvers*, not as information processors. This is not the norm of call centre organising, as the body of critical literature has established. But why should call centre work be routine, production work? Can the fundamental assumptions be questioned by those managers with the deepest understanding of their shortcomings?

It is also clear that a key dynamic in call centre operation is the degree to which management practices are centrally prescribed, leading to an erosion of the responsive autonomy of local managers. Whereas managers at Quotes Direct had little sense of influence over definitions of activity, at Housing Helpline, autonomy was a prized feature of the environment. However, the keen visibility of performance in the call centre compared with a more conventional form of organising meant that even the protective environment of Housing Helpline remained vulnerable.

> I do miss having a big staff – its easier to manage 100 plus staff than to manage the 40 odd we've got here because you've more resilience to things. Here if we get a queue in the morning at 9am, we can't recoup that, we just can't do it. If I take a hit in a big call centre at 9am I just say to them 'delay all your breaks, do this, do that . . .' and with that many people on the floor there is more experience and knowledge so you can still take people off the phones. If you take one off here sometimes, it kills the floor.
>
> (Barbara, Call Centre Manager – Housing Helpline)

So while the autonomy to manage the Housing Helpline centre in an open and flexible way was partly protected by its small size, this also meant greater exposure to demand and service level fluctuations, making the targets they were committed to a real challenge. Partly in response to this, Housing Helpline began to act as a service provider to other organisations, thus leveraging their resources and creating a new income stream (which allowed them to increase staff numbers). However, the externally imposed performance metrics and cost drivers associated with such third party business pose a further threat to Housing Helpline's espoused value system around quality, service and call handler roles. In these conditions, the pressure to maximise capacity and productivity through more conventional call centre management practice comes to the fore.

The review of these cases points to the tensions between control and empowerment in the call centre. However it suggests that call centre managers are driven not by 'controlling intentions' (Thompson and Ackroyd 1995), but rather by a desperate struggle to balance contradictory demands. The voices of managers in this chapter illustrate the precarious, dependent and uncertain nature of management. For now, the balance appears to work at Housing Helpline, but the tensions and tendencies that drive Quotes Direct are not far from the surface.

The managers included in this study could be described as active role constructors who make choices and exercise agency. Their accounts present evidence of considerable innovation, improvisation and variety in their practice. However they do so within a highly constrained context of goal contradictions, system dominance and pressure to normalise that appears to take away many of their choices. Ultimately therefore, while frontline managers in the call centre are key holders of knowledge concerning the complexity and contradictions of the work they supervise, in their consignment to micromanagement, they are both silenced and distanced from influence. The conclusion to be drawn is that rather than being over-managed, call centres tend towards a shortfall of strategic leadership.

By exploring the subjective experiences of these frontline managers, this chapter has offered a perspective that is substantially missing in the call centre literature. Call centre management is typically portrayed as autonomous and deterministic and yet a picture has emerged here of managers who creatively innovate, but nevertheless struggle with conflicting role requirements and contextual constraints and often lack the support they need.

The challenge of achieving efficiency and providing customer service, while withstanding strong pressures towards a cost and control model, calls for managers with nuanced understandings of the realities of the frontline, who also have strategic influence. Further research and action can usefully address the lines of strategic support for frontline managers in order to better influence the future prospects and challenges for this sector.

Notes

1 An earlier version of this chapter appears as: Houlihan, M. (2001) 'Managing to Manage? Stories from the Call Centre Floor', *Journal of European Industrial Training*, 25: 208–221.

2 Because Quotes Direct represents a fairly typical call centre, I give a more detailed account here of Housing Helpline's formative context.
3 Agent's PCs at Housing Helpline had all the facilities of a normal PC including word processing and email, in addition to the customised housing management platform, in common with regular employees at Housing Association. Thus agents wrote internal emails or letters to external parties to action particular customer requirements as part of the normal process of conducting their work. At Quotes Direct, agents' computers were the more typical 'dumb terminals' which enabled only the encoded call management and associated database completion procedures.

References

Batt, R. (1999) 'Work Organisation, Technology and Performance in Customer Service and Sales', *Industrial Relations and Labour Review*, 52(4): 539–562.

Belt, V., Richardson, R. and Webster, J. (2002) 'Women, Social Skill and Interactive Service Work in Telephone Call Centres', *New Technology, Work and Employment*, 17(1): 20–34.

Callaghan, G. and Thompson, P. (2002) 'We Recruit Attitude: The Selection and Shaping of Routine Call Centre Labour', *Journal of Management Studies*, 39(2): 233–254.

Cameron, D. (2000) *Good to Talk?*, London: Sage.

Dalton, M. (1959) *Men Who Manage: Fusions of Feeling in Theory and Administration*, New York: John Wiley and Sons.

Frenkel, S. J., Tam, M., Korczynski, M. and Shire, K. (1998) 'Beyond Bureaucracy? Work Organisation in Call Centres', *International Journal of Human Resource Management*, 9(6): 957–979.

Frenkel, S. J., Korczynski, M., Shire, K. and Tam, M. (1999) *On The Front Line: Organisation of Work in the Information Economy*, USA: ILR and Cornell University Press.

Hales, C. (1999) 'Why do Managers do What they do? Reconciling Evidence and Theory in Accounts of Managerial Work', *British Journal of Management*, 10: 335–350.

Houlihan, M. (2001) 'Managing to Manage? Stories from the Call Centre Floor', *Journal of European Industrial Training*, 25(2): 208–220.

Houlihan, M. (2002) 'Tensions and Variations in Management Strategies in Call Centres', *Human Resource Management Journal*, 12(4): 67–86.

Jackall, R. (1988) *Moral Mazes: The World of Corporate Managers*, New York: Oxford University Press.

Kanter, R. (1979) 'Power Failure in Management Circuits', *Harvard Business Review*, 57(4): 65–74.

Kinnie, N., Hutchinson, S. and Purcell, J. (2000) 'Fun and Surveillance: The Paradox of High Commitment Management in Call Centres', *International Journal of Human Resource Management*, 11(5): 967–985.

Knights, D. and McCabe, D. (1998) 'What Happens when the Phone goes Wild? Staff, Stress and Spaces for Escape in a BPR Telephone Banking Work Regime', *Journal of Management Studies*, 35(2): 163–194.

Korczynski, M. (2001) 'The Contradictions of Service Work: Call Centre as Customer Oriented Bureaucracy', in A. Sturdy, I. Grugulis, and H. Willmott (eds) *Customer Service: Empowerment and Entrapment*, Basingstoke: Palgrave.

Korczynski, M. (2002) *Human Resource Management in Service Work*, Basingstoke: Palgrave.

Lankshear, G., Cook, P., Mason, D., Coates, S. and Button, G. (2001) 'Call Centre Employees' Responses to Electronic Monitoring: Some Research Findings', *Work, Employment and Society*, 15(3): 595–605.

Leidner, R. (1993) *Fast Food, Fast Talk: Service Work and the Routinisation of Everyday Life*, Berkeley: University of California Press.

Taylor, P. and Bain, P. (1999) 'An Assembly Line in the Head': Work and Employment Relations in the Call Centre', *Industrial Relations Journal*, 30(2): 101–117.

Taylor, P., Mulvey, G., Hyman, J. and Bain, P. (2002) 'Work Organisation, Control and the Experience of Work in Call Centres', *Work, Employment and Society*, 16(1): 133–150.

Thompson, P. and Ackroyd, S. (1995) 'All Quiet on the Workplace Front? A Critique of Recent Trends in British Industrial Sociology', *Sociology*, 29(4): 615–633.

Wallace, C. M., Eagleson, G. and Waldersee, R. (2000) 'The Sacrificial HR Strategy in Call Centers', *International Journal of Service Industry Management*, 11(2): 174–184.

Willmott, H. (1997) 'Rethinking Management and Managerial Work', *Human Relations*, 50(11): 1329–1350.

10 How 'Taylorised' is call centre work?

The sphere of customer practice in Greece

Aikaterini Koskina

Introduction

Despite the global presence of call centres, most of the academic contributions to date are US/UK driven and are framed within a 'contingency' theoretical perspective,[1] where technology plays the most influential role in the call centre labour process. Whilst this is valuable in its own right, little emphasis has been placed on 'institutional' comparative analysis, especially within countries where call centres have recently started to attract public attention. This chapter begins to address this research gap by examining whether technology does, in fact, play the most influential role in guiding the nature and management of call centre work within a selection of organisations operating in Greece.

It has been argued that call centres are at the forefront of increased automation, with close monitoring, tight control, high routinisation, labour division and strict quality adherence occurring through the integration of telephone and computer technologies (Arkin 1997; Taylor and Bain 1999; Callaghan and Thompson 2001; Bain *et al.* 2002; Deery and Kinnie 2004). The underlying premise here is that call centre environments are an extension of 'Taylorist' approaches – 'the call centre labour process represents new developments in the Taylorisation of white collar work' (Taylor and Bain 1999, p. 115). In this context, as featured in Figure 10.1, employees are perceived as a measurable entity with identifiable physical and mental traits (Sawyer 1989) that need to be manipulated by management to fit the requirements of production (Watson 2003). 'Taylor[ism] places emphasis on the role of the worker as a general-purpose machine operated by management' (Braverman 1974, p. 124). Call centre workers, therefore, appear to occupy 'low-discretion' and 'low-trust' roles, in which management holds a high level of 'technical', 'detailed' and 'bureaucratic' control.

According to Edwards's (1979, p. 21) threefold typology of control, *direct or simple control* mainly exists in small firms and implies the exercise of personal power by the entrepreneur and the absence of formal rules. Where *technical control* occurs, machinery sets the pace of work, and where *bureaucratic control* occurs, it relates to the institutionalisation of hierarchical power. Another important analytical distinction here is the control that transpires between detailed and general control. *Detailed control* refers to the details of work tasks, whereas *general control* covers the accommodation of workers to the overall aims of the enterprise (Edwards 1986, p. 6). This is consistent with Fox's (1974, pp. 19–20) model of trust and work roles. A central tenet of his theory is that no role can be totally diffused or totally specific, thus, all jobs involve decision-making. Trust is concerned with relationships, which are structured and institution-alised in the form of roles and values. *High-trust* and *high-discretion* work patterns involve a high degree of moral involvement and employment relations are characterised by open-ended exchanges with a high degree of employee commitment. In *low-discretion, low-trust* positions, management has no need to trust its employees because of the detailed division of labour whilst both parties act in a cautious calculative manner. In these work roles, employees are seen as atomistic individuals who can be slotted into technological and bureaucratic systems, as happens within the 'Tay-lorised' call centre environments presented earlier by existing studies.

Conversely, there is evidence from the literature on frontline work and an increasing body of work from cross-national studies which suggests that management and employment systems can be nation, sector and occupa-tion specific. The first set of these studies (Leidner 1996; Frenkel *et al.* 1998; Taylor *et al.* 2002) proposes that variations of control can be related to differences in the discretionary content of work roles. In the cross-national evidence (Hofstede 1980; Whitley 1992; Marginson and Sisson 1994; Guest and Hoque 1996; Ferner 1997; Ferner and Quantanilla 1998; Katz and Darbishire 2000) lies the argument that the national identity of the corporate governance (country of origin effect) and the cultural insti-tutions of the host nation (host country effect) can influence the nature and management of work. Indeed, this ideology fits well within the

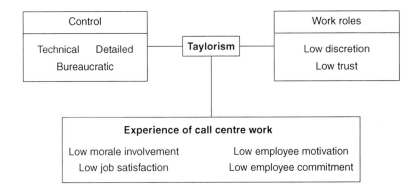

Figure 10.1 Taylorist model of call centre work.

institutional approach, which sees 'the business organisation as socially constituted and thus reflecting national distinctiveness in institutional arrangements' (Lane 1989, p. 31). Clearly, this is something that the contingency school either fails to recognise or explicitly rejects (Child and Tayeb 1982–1983).

The aim of this study is to test the assumption of the contingency perspective, whereby it is claimed that the role of technology and, in turn 'Taylorism', is a driving influence for the convergence in call centre work. For the purpose of this chapter, this postulation has been termed the *call centre diffusion thesis*. The call centre labour process in the organisations studied is examined within an institutional framework, where comparison focuses on the interplay between organisational processes on the one hand and different industrial settings on the other. The first part of the chapter outlines the methodological approach of the study, while the second draws out the research findings in terms of three dimensions: organisational structures, work organisation and performance management. The emphasis here is upon management approaches and the structure of work that sets the context in which to explain the experience of call centre work in Greece. The final part of the chapter focuses on the wider implications of the findings for call centre employees, employers and policy makers across national borders.

Research sites and methodology

The call centre sector is a growing industry in Greece, with a focus on the areas of telecommunications, banking, insurance and tourism, where over 3,000 employees are currently employed, and where 'no major research has yet been carried out' (EIRR 2000, p. 13).

The research was conducted in Athens. This was because the city's employment in services in general (Kritsantonis 1998) and in call centres in particular (EIRR 2000) is expanding rapidly, predominantly in the telecommunications and insurance industries. Telecommunications in Greece is a new and developing sector. It is estimated that, today, approximately 15 corporations operate in the telecommunications market, of which the vast majority has a Greek ownership (*Epilogi* 2003). On the other hand, the insurance industry, despite being a long-established market within the Greek business environment, has expanded rapidly since the late 1980s, with an annual average growth rate of 20 per cent (*The Economist* 2001). At present, nearly half of the 69 insurance companies operating in Greece are foreign-owned (*Asfalistiki Agora* 2004).

Empirical evidence was collected from four call centres between May and October 2004. These call centres provide a wide range of customer services (see Table 10.1). The sites were matched as far as possible in terms of call centre environment, size and nationality of ownership. Two workplaces are from the telecommunications industry and two from insur-

Table 10.1 Profile of case study organisations

Organisations	Sector	Market share[c]	Annual profit	Number of employees[d] 50–100	Number of employees[d] 100+	Call centre workers (%)[e]
Foreign A[a]	Telecommunications	41.05	Confidential		✓	40
Greek A	Telecommunications	1.40	Confidential	✓		90
Foreign B[b]	Insurance	0.80	€15 mil.	✓		30
Greek B	Insurance	2.60	€48 mil.		✓	60

Notes
a Foreign A operates under European ownership.
b Foreign B operates under German ownership.
c Share of the Greek telecommunications and insurance market.
d The exact figures of employees are confidential.
e Call centre workers as percentage of each company's total workforce.

ance services. In both sectors, one case study concerns a Greek-based firm and one a foreign-owned workplace that is operating in Greece. Each of the participant centres is non-unionised and employs between 80 and 250 operators who deal with a high volume of inbound and outbound calls.

The 'Foreign A' case study is part of a leading mobile telephony provider that has been in the Greek market for nearly 11 years. The 'Greek A' call centre was established in late 1999 and belongs to a high growth company providing landline telephony and Internet services. The insurance case study organisations, which are currently ranked amongst the top 20 best performing units in the insurance field, used to be family-owned businesses. The 'Foreign B' case study has operated under German ownership in the Greek market since 1997 and specialises in home and car insurance. The 'Greek B' workplace, which is a multinational, was established in 1953 and provides general insurance and financial services.

In terms of methodology, the research involved extensive semi-structured interviews with different levels of management in the four call centres and a self-administered, semi-structured questionnaire of the call centre workers in three case study organisations (one could not participate). The survey was based on a common design so that causal mechanisms could be explored. The first set of questions aimed to elicit background information about the workplace and the job of the participants. The second set looked at performance management. The third set of questions focused on the attitudes of the participants towards their work and management. A range of documentary evidence, including company policies, reports, newsletters and statistical data, were also examined. In addition, a random sample of call centre workers in the Greek-owned organisations was interviewed as a supplementary source of qualitative data. Table 10.2 gives an overview of the research undertaken in the four call centres.

In Foreign A, management did not allow employees to participate in the survey due to the heavy workload of preparation and support for the Olympic/ParaOlympic Games held in Athens during the period in which the fieldwork was conducted. In the other three call centres, management did not allow the entire workforce to be interviewed. Therefore, interviewees (employees) were chosen at random and were invited to participate in the survey. Unlike the limitations on the number of interviews, the questionnaire was distributed across the total call centre workforce.

The sphere of call centre work in the case study organisations

The research investigated the extent to which the Greek call centre labour process is 'Taylorised'. The emphasis here is upon three entrepreneurial dimensions of call centres: organisational structures, work organisation and performance management. This sets the context in which to examine the boundaries of 'Taylorism', as outlined in Figure 10.1 (control, work

Table 10.2 Research methods and participation rates

| | Questionnaires | | Interviews | | Documentary evidence |
	Employees (n=69)	*Respondents (%)[a]* (n=69)	*Employees* (n=11)	*Management* (n=41)	(n=18)
Foreign A	✓			✓	
Greek A	✓	60	✓	✓	
Foreign B	✓	25	✓	✓	✓
Greek B	✓	70	✓	✓	✓

Note
a Respondents as percentage of each organisation's call centre workers.

roles, experience of work). In the following analysis the dimension of organisational structures has been used to identify the degree of bureaucratisation, whereas the other two dimensions have been used to measure the degree of technical/detailed control and to portray the experience of work. Table 10.3 provides a summary of the converging–diverging elements of the four organisational dimensions examined, which are covered subsequently in this section. The evidence of the case studies presented here demonstrates that the Greek call centre workers involved in this study occupy work roles in which the roles of different forms of control are far distant from 'Taylorist' traditions (see Table 10.4).

Table 10.3 The call centre labour process in the four case studies

Telecommunications industry		Insurance Industry	
Foreign A	*Greek A*	*Foreign B*	*Greek B*
Organisational structures			
Vertical structure	Vertical structure	Vertical structure	Vertical structure
Flat hierarchy	Flat hierarchy	Flat hierarchy	Flat hierarchy
Decentralisation[a]	Centralisation	Decentralisation	Centralisation
200–250 operators	80–100 operators	30–40 operators	100–120 operators
4 supervisors	4 supervisors	No supervisors	4 line managers
			4 supervisors
Work organisation			
General control		General control	
No transcript (except billing section)		No transcript	
Autonomy/satisfaction		Autonomy/satisfaction	
No monitoring		No monitoring	
Performance management			
No qualitative/quantitative targets (except billing section)		No qualitative/quantitative targets	
Emphasis on quantity		Emphasis on quality	
Written report: weekly		Oral report: fortnightly/monthly	
No financial incentives	Ex gratia non-profit related bonus every 4 months	No financial incentives	Twice a year: €500–700 bonus
Family-friendly climate	Trust-based culture	Family-friendly climate	Monthly: take employees out
Relaxed atmosphere		Relaxed atmosphere	Annually: 2-day trip
Trust-based culture		Trust-based culture	Trust-based culture

Note
a Centralised/decentralised refers to the decision making approach followed in each call centre regarding the organisation of work, employee development and performance management.

Table 10.4 The 'non-Taylorisation' of call centre work in the four case studies

	Telecommunications industry		Insurance industry	
	Foreign A[d]	Greek A	Foreign B	Greek B
Control[a]				
Technical	Low	Low	Low	Low
Detailed	Low	Low	Low	Low
Bureaucratic	Low	Low	Low	Low
Work roles[b]				
Discretion	High	High	High	High
Trust	High	High	High	High
Experience of work[c]				
Morale involvement	–	High	High	High
Job satisfaction	–	High	High	High
Employee motivation	High	High	High	High
Employee commitment	High	High	High	High
Performance monitoring	Low	Low	Low	Low

Notes
a Evidence is based on survey/documentary observations.
b Evidence is based on survey observations.
c Evidence is based on survey/documentary observations.
d Evidence is based on management responses.

Organisational structures

The organisational structures of the overseas-owned case study organisations were somewhat similar, despite the heterogeneity of their size and
industry. Both companies possessed a formal conventional organisational
structure, which included more than eight main divisions and a separate
customer care unit. Despite the highly centralised decision-making structure at the group level, individual departments were predominantly decentralised. In Foreign A, customer services engaged between 200 and 250
operators. They operated 24 hours a day for seven days a week and
employees were overseen by four team leaders. In Foreign B, more than
25 operators provided customer care for eight hours per day during week-
days. Two senior German Vice-Presidents were responsible for all divisions and there were no layers of middle management or supervisors
within the different sections – including customer care. Indeed, this vertical organisational pattern, according to the German directors in Foreign B
and the Greek team leaders in Foreign A, provided commitment that built
credibility and motivation behind the corporate strategy. It enabled a
linear relationship between management and employees and a homogeneous pattern across different divisions. In this way, it was much easier
to assimilate the mother companies' organisational framework with the
prevalent one in the host country.

With the Greek-owned organisations, centralised decision-making did not take place at the group level but at the departmental level. Both customer service units were subdivided into four operations where between 25 and 35 frontline staff work the same hours and days as Foreign B. Although there is a single company-wide HR policy, the various sections maintain independent HR practices according to the needs of their employees. Each division was essentially run as a separate unit and was administrated by one line manager and one supervisor in Greek B and one supervisor in Greek A. The prime rationale for this structure, according to one of the General Managers, was to 'make things function better since it enables individuals in different areas to agree on a common goal, and as a result, the employees co-operate with one another rather than compete'.

Evidently the converging element between these four enterprises was that they are all based on vertical hierarchical structures. The point here, for the insurance companies, was that the simple transfer of corporate ownership had not dramatically altered the organisational structures of these environments, as their management processes remained deeply rooted within the old paradigm of the family-owned business. However, in the case of the telecommunications environments, their management structure appeared to be influenced by the traditional Greek management framework, where patriarchal and individualistic norms occur. On the divergence side, in the Greek-owned call centres, decision-making appeared to be centralised ('top-down'), whereas in the non-Greek call centres, decision-making was found to be decentralised ('bottom-up'). Nevertheless, this background analysis suggests that, on average, there was one supervisor for 30 advisors and as a result, it would appear that a relatively low level of surveillance took place.

Work organisation

The call centre workers in both industries worked within open-plan offices and used computer terminals in order to locate, store or amend any customer-related information and inquiries. The operators' main role was to maintain regular contact with customers by telephone, email, fax or post. The vast majority of the respondents in the telecommunications sector were male workers aged between 20 and 29 years old, who had been employed by the case study organisations for between one and two years. In the insurance industry, the average age of the female employees, who account for more than half of the participants, was between 30 and 39 years old and they had held their current positions for more than five years. Both sectors offered full-time jobs on open-ended contracts. These contracts were normally between 35 and 40 hours a week and involved shift work in telecommunications and overtime work in insurance. The average salary of call centre workers was between 9,000 and 13,000 per year, including medical insurance, pension scheme contributions, allowances for holidays (equivalent to 25 working days) and sick leave. In addition, all Greek workers, including operators, according to Greek labour legislation,[2] are entitled to 'epidomata', which involves extra disbursements of a full month's wage during Christmas and a half-month's wage over Easter and summer.

Within telecommunications, the advisors were grouped into four teams: general inquiries, contracts/pre-paid packages, technical support and billing. The first three teams were responsible for handling inbound calls relating to issues such as accessing and updating customers' records, dealing with general and directory inquiries, complaints, and providing advice about products and services. The work of the billing team, which was subdivided into company and personal accounts, involved mainly dealing with outbound calls regarding outstanding payments. The workers in the billing section used a transcript during their first days at work until they felt confident to conduct their conversations independently with the customers. With regard to the remainder of the advisors, they utilised an electronic system that included a customer database and a printed manual with all the products and services provided by these companies. In both divisions, there was no direct control or supervision over their work. When one of the operators in Greek A was asked whether he felt trapped in this workplace, he answered:

> That's a rather funny question you're asking ... I don't see why I should feel that way ... I am not afraid to leave my desk ... when I need to go to the toilet or anywhere else I just go ... if I'm having a problem with a call and I cannot deal with it then I pass it to the supervisor ... that's the point of having somebody more senior in this room after all.

Likewise, the customer advisors in the insurance sector appeared to have discretion in the design and provision of services. Overall, they handled claims, either directly with individual/corporate customers, or through the independent agents who supported the firm's sales system. There was neither an electronic nor a hard copy of a transcript that they needed to follow during their telephone contact with the customers. The only tools that they had to use, according to a Greek line manager, were 'their intelligence, confidence, expertise, knowledge and diplomacy'. In addition, as one advisor in Greek B suggested, 'self-discipline' is seen as one of the most important skills in this work because it enables workers to manage challenging calls and to cope with frustrating customers. As with the telecommunications case, there was only one supervisor in each division, who was located in a separate room from the operators, and whose main role was not to 'safeguard' the agents but instead to provide assistance and support when customer-related problems arose. As one agent noted, the level of surveillance and control was extremely weak.

> We are given lots of freedom and autonomy ... I'm my own boss ... I handle each call according to my personal preferences, such as the pace I speak at and most importantly how I deal with each inquiry ... I think that's great, as they trust us with what we do ... they have faith and confidence in our work and that makes you try even harder to do your work as well as possible.

Within the call centre environments described, the frontline workers were found not only to occupy high-discretionary work roles but also to have a great sense of satisfaction in their work.

> I find my job challenging and stimulating ... I deal with a variety of issues, from consultation to representation, and with different types of customers, from neurotic to extremely nice ones ... there's plenty of opportunity here to add value to your work, if you want to, the more you do it, the more you learn, and the more you like it. Definitely, I wouldn't exchange my job with typical boring office work even if I were offered a slightly higher salary.
>
> (Operator in Greek B)

> I'm quite happy working here ... I like the idea of dealing with people and trying to find a solution to their problems ... I prefer this than selling things like I did in my previous job. Everything here is relaxed and even if I get boring calls, like changes in customers' details, that's only a small part of my job that gives me a chance to get a break from customer complaints.
>
> (Operator in Greek A)

Clearly, in both examples, the work of these operators is not attuned to follow a predetermined set of rules laid down by management. Under these conditions, the degree of autonomy employees possessed over the provision of services across the three centres led to a common sense of job satisfaction.

Performance management

The performance review systems in Foreign A, Foreign B and Greek A were company-wide, whereas in Greek B they were department specific. Each multinational corporation's subsidiary implements formalised policies that are designed at the corporate headquarters and they were implemented in a less formalised manner according to the needs of each division. Typically, in the case of telecommunications, the operators met with their team leaders on a weekly basis. In Foreign B, however, meetings took place every fortnight, and in Greek B monthly. All workers have to prepare a report on the number and nature of calls dealt with. In the insurance companies this report is oral.

Within the four settings, the primary focus was on the provision of quality service rather than on tight control and high surveillance. In the insurance sector case studies, any specific targets were non-existent. As the manager in Greek B explained,

> Calls may vary from a couple of seconds to over half an hour ... it depends on the situation ... each call is different, the question is different and the answer is not always that simple. There is no script and we do not want to introduce such a thing, as our people and our customers are human beings who need to communicate ... For example, when there is a fatal car accident the situation is rather complicated as on the one hand we have to make a claim on behalf of our client and on the other we have to deal with the other party who has lost a relative ... a human life ... and we certainly cannot simply forget about the grief of these people ... our job is rather difficult and frustrating ... our principal aim is to sustain a 'human face' both for our employees, our customers and our competitors and not to meet quantitative targets ... that's our strength!

In contrast, in the telecommunications industry, although there were no restrictions on the length of calls and no fixed quantitative targets, except for outbound contacts relating to overdue accounts, there was the perception that quantitative numbers were much preferred. Though these companies presented various factors on the list of their priorities, one point kept recurring in interviews with managers. This was that efficiency was really important when operators dealt with such issues as directory enquiries and updates of customers' personal details. Indeed, when advisors had to manage more complex situations, like complaints, software enquires and outstanding balances, quality concerns were more important. The following claims, made by an advisor, provide a clearer picture of the current situation in Greek A.

> My main responsibility is to ensure that calls are covered all the time. How quickly you handle each call really depends on your experience and not on management's pressures. For instance, during my first month here it took me over 40 minutes to activate a connection, whereas now I can do it within less than 10 minutes.

Each of the two Greek employers had implemented an 'informal tradition', as one of the supervisors suggested, to 'reward their employees for their hard work'. It should be noted that none of these companies had a performance-related payment system and that these financial incentives were offered in addition to the 'epidomata' the operators receive. In Greek A, every four months all employees received a bonus, which was *ex gratia*, and not profit related. In Greek B, management took all the employees out at the end of each month, and on an annual basis offered them the opportunity for a two-day trip with all the expenses covered by the company. In addition, twice a year each advisor received a bonus of 500 to 700, which is almost equivalent to two month's wages. When one of the managers was asked why there is such a tradition, he replied:

> These people are the reason for our success . . . their work has kept the company in such good financial shape all these years . . . without their spirit none of this would have ever been possible . . . this is the least that we can do in order to thank them.

Conversely, the non-Greek call centre employers, despite not providing any financial incentives, adopted a strategy of developing a 'family-friendly' culture within the workplace, which assisted in motivating their staff. The overall aim of this approach was to make the advisors feel comfortable talking to their superiors about any work-related problems they might have. In Foreign B, the Vice-Presidents tried to discuss non-work-related issues with employees (such as their plans for the weekend) on a daily basis, whereas in Foreign A, there was a tendency for a relaxed environment where hierarchies were of little importance.

> From day one I try to assure all the advisors I am on their side . . . I am their friend and not their enemy that they need to be afraid of . . . Of course, this needs to be developed over time and up to now it seems to work well . . . being on the phone for eight hours a day sometimes can be really stressful so when lines are extremely busy I'm taking many of these calls on top of my workload in order to decrease a bit their work and of course to keep as many customers as possible satisfied.
>
> (Team Leader in Foreign A)

Certainly, this evidence suggests that across the four call centres there was a requirement for disparate performance management schemes to accommodate the individual needs of both employees and the organisation. Despite the variations in the style of performance measurement and employee motivation strategies followed in each of the four workplaces, there was an overall convergence in the emphasis placed on the creation and maintenance of a 'trust-based' culture.

So, is there a Greek approach to call centre management?

The evidence, presented in this chapter, suggests that although the nature and management of call centre work in Greece is embedded in consensus concerning organisational values and goals, it emerges from unconstrained dialogue among organisational members. As a result, it is not always promulgated just from the top. Authority within the four case study organisations was informal and focused on individual influences rather than on hierarchical disparities. Moreover, the entrepreneurial environment of the four call centres tended to encourage the development of a network of informal relationships that cut across hierarchical boundaries. Although, in theory, there was a clear formal vertical structure, in pragmatic terms, the informal employment mechanisms in these workplaces assisted in the development of a flatter hierarchy, where managers possessed 'general control' without maximising 'detailed control'. In Foreign B, for instance, the flattening of hierarchies was mentioned in the same breath as 'informal teaming'. Surprisingly, this management approach was enforced, in a sense, by the distinctive cultural dimensions of the Greek workforce and not by the German ownership of this workplace.

> The German culture is characterised by rigid and inflexible working patterns. For instance, it would not be possible for an employee in the headquarters to leave earlier or arrive later at work without, first, notifying the line manager at least two days in advance and second, without replacing the minutes of work lost. Undoubtedly, a similar system could not work effectively in this environment. Greek workers are more difficult to manage than German ones. They are less likely to obey a set of rigid rules and they always have the need for flexibility and informal relationships. Coming from a German background it took me at least a year to realise this. Once I tried to adapt to the local environment by allowing employees greater discretion in their work, greater flexibility regarding working time and a more relaxed atmosphere without the intervention of middlemen, such as supervisors, this led to a completely different organisational climate. Most of the workers were highly productive, collaborative and, most surprisingly, willing to work overtime when required without extra pay, or even accepted to undertake responsibilities and work outside their job boundaries. The latter is something that you cannot find easily in the German workforce, or at least with those employees with whom I have worked for over fifteen years. This is what I call informal flexible team working, which adapts to the individual needs of the organisation and the employees.
>
> (Vice-President in Foreign B)

These comments perhaps directly confirm the argument that 'the labour process contains within it an area of uncertainty which reflects the fact that no contract can specify workers' duties in exact detail' (Edwards 1986, p. 7). Though the present study has not focused on the patterns of conflict and control in call centre environments, the picture emerging in Foreign B indicates that if 'detailed' control is evident in the bargaining agenda between employees and managers, the problem of securing compliance arises. Then utilitarian forms of worker resistance are likely to take place, since these environments are individualised and non-unionised. Perhaps, this situation fits well with the 'militant' but 'individualised' orientation of Edward's (1986, pp. 231–233) model of workplace relations.[3] In this pattern, the characteristics of an industry tend to create certain orientations among its workers, which, in turn, encourage a common response: the means whereby this response is practised becomes an *interest* for the workers and is *valued* by them. Indeed, this is a special case and it has been somewhat neglected, as it has generally been assumed that militancy and collectivism necessarily go together. Thus, the evidence provided here cannot support such an argument, it provides a possible direction for future research.

The central point here is that both the structure of managerial authority and the pattern of relations between managers and employees reflect a 'loose–tight' relationship that is founded on the 'moral involvement' of the employees (Fox 1974) or in Greek terms, on the 'philotimo' of the indigenous workers. Broome (1996, pp. 66–69) describes this fundamental characteristic of the Greek mindset as a complex set of values and attitudes involving honour, obligation, self-esteem and appropriate behaviour towards members of one's in-group. Indeed, 'the characteristics of successful firms involve simultaneous loose–tight properties [which] are on the one hand rigidly controlled yet at the same time allow maximum individual autonomy' (Edwards 1987, p. 90). In the Greek business environment, conceivably the pivotal concept is not to differentiate the individual from the workplace but instead to place emphasis on the individual needs of employees as they arise. A salient feature in harmonious employee relations and high performing environments in Greece is 'the ability to treat each employee as a person' (Papalexandris 1999, p. 44). Greek workers might be difficult to manage, as the German director in the insurance company indicated, but 'being an effective manager in Greece can be both very challenging, as you need to develop flexibility and gain deeper understanding of the employees' individual needs, as it is very difficult if you ignore these issues' (Papalexandris 1999, p. 57). In other words, as Broome (1996, p. 79) suggested, 'in Greece you must manage persons, not personnel' and certainly, as this study indicated, this is the case for call centre environments as well.

These peculiarities partly explain the rather strange, conflicting, but at the same time, complementary management style of the case study organisations, which assert both autocratic ideologies and egalitarian values without the presence of trade unions. It is true that 'control is not simply imposed by management but emerges as an amalgam of different elements' (Edwards 1986, p. 41). In the Greek example, managerial prerogatives are shaped and influenced by the distinctive cultural traits of the workers, which are tightly interwoven with the political history of the country. The long periods of various occupations and dictatorships have assisted in developing a high tendency amongst Greeks towards individualism, autonomy, internal control and self-esteem (Cummings and Scmidt 1972; Broome 1996; Papalexandris 1999). Within the Greek entrepreneurial context, perhaps the central aspects of a well-balanced employment relation are 'eleftheria', which means freedom, and a strong sense of 'philotimo' (Broome 1996, p. 66).

In the co-existence of managerial control and employee autonomy lies 'the conception of consensual legitimacy which is known as democracy' (Heckscher 1994, p. 39). In Greek terms, this relates to the power of people at work, since 'demos' implies people and 'kratos' entails strength. Whether power resides at the top or at the bottom of the pyramid, however, when monopolistic and monopsonistic pressures might enforce unilateral management control over quantitative performance targets, is not known. The evidence presented in this chapter, which represents a beginning for the exploration of the terrain of the call centre labour process in Greece, has focused on the internal organisation of work. Thus, further examination is required of the external environment of call centre work in Greece.

The paradox of the call centre diffusion thesis

The image that emerges here highlights the non-conformist and idiosyncratic character of the Greek call centre labour process. Call centre work in both foreign-owned and indigenous workplaces within the specific national setting appears to be disengaged from systems of detailed, bureaucratic and technical control. The 'loose–tight' employment relationship is strongly associated with work autonomy, job satisfaction, and employee motivation–retention–commitment, but not necessarily with low levels of work intensity. Fox (1974) stresses that 'high-trust' work roles, as in this study, are linked to 'high-discretionary' work content. In this sense, 'detailed supervision is considered to be inappropriate because the control comes from within – it is, in a literal sense, self-control, and the emphasis is on joint problem solving rather than on the unilateral imposition of procedures' (Fox 1974, p. 20). Undoubtedly, therefore, the character of Greek call centre employment is markedly distant from the omniscient Taylorist accounts and confirms the point that in none of the four organisational settings was the management and organisation of work driven by the use of technology.

To discuss call centres is to occlude the fact that even these customer-oriented and technologically intensive environments, which operate within contemporary capitalist societies, are complex systems of mutually dependent individuals. The Greek example of this research clearly indicates that the overall call centre labour process appears to be more of a divergent phenomenon, which is likely to be influenced by the institutional characteristics of a nation rather than just a consequence of technological driving forces. In the rhetoric of the call centre diffusion thesis is envisioned a working pattern in which technology can somehow overcome the complex heterogeneity of host countries' national distinctive traits. Again, in many contexts, such as in the large call centres located in Anglo-Saxon countries, Taylorist claims may have theoretical plausibility. Though the host country of this research has always been and still is the realm of small–medium enterprises (Kritsantonis 1998), the case study organisations, or at least two of them according to Greek norms, are classi-fied as large employers. In agreement with Fox (1974 p. 23), 'in the large firm as in the small, much white-collar work is non-repetitive and requires a modicum of responsibility and individual judgment', as can be seen in the four call centres.

Since the principal conclusion of this study suggests that the host country national context exerts a much stronger influence on the character and effectiveness of the call centre labour process than technology, the universality of the contingency propositions should be questioned. Within cross-national comparative studies it has been argued that 'national differ-ences in organisations, rooted in national institutions, casts doubt on the notion of a single type of modern bureaucracy governed by universalistic standards and molded by common contingencies' (Child and Tayeb 1982–1983, p. 48). The question within such studies should focus on whether the disparate uses and applications of technology can influence the management and employment practices of call centre work rather than how technology determines these systems. In this context, a primary task for academics, employers and policy makers is to enquire whether the par-adigms of the call centre diffusion thesis might have an ideological subtext within different national contexts.

Acknowledgement

I am very grateful to Dr Carole Thornley for the support, enlightenment and encouragement. Thanks to Dr Steve French for advice on the devel-opment of this study. Thanks also to Steve Funnell, Julia Connell and John Burgess for comments on an earlier draft of this chapter. Finally, it should be noticed that this research project, which forms part of my PhD thesis, was funded by the School of Economic and Management Studies at Keele University, UK.

Notes

1 A main essence of the 'contingency perspective' lies in the argument that 'technology determines organisational structure and behaviour and that the resulting organisational characteristics will be stable across nations, regardless of any differences between industrial nations in culture or forms of ownership of productive resources' (Lane 1989, p. 22).
2 Greek Labour Directive: 1082/1980.
3 In Edwards's (1986, p. 226) model, 'militant' refers to the extent to which

workers perceive themselves as having interests which are inconsistent with the interests of management and act accordingly. 'Orientation', which can be individualised or collectivised, in this context means an approach which influences behaviour within the workplace.

References

Arkin, A. (1997) 'Hold the Production Line', *People Management*, 6(3): 22–27.

Asfalistiki Agora (2004) 'Performance of Insurance Companies', September, pp. 50–51 (in Greek).

Bain, P., Watson, A., Mulvey, G., Taylor, P. and Gall, G. (2002) 'Taylorism, Targets and the Pursuit of Quantity and Quality by Call Centre Management', *New Technology, Work and Employment*, 17(3): 170–185.

Braverman, H. (1974) *Labour Monopoly and Capital*, New York: Monthly Review Press.

Broome, B. J. (1996) *Exploring the Greek Mosaic*, Yarmouth: Intercultural Press.

Callaghan, G. and Thompson, P. (2001) 'Edwards Revisited: Technical Control and Call Centres', *Economic and Industrial Democracy*, 22(1): 13–37.

Child, J. and Tayeb, M. (1982–1983) 'Theoretical Perspectives in Cross-National Organisational Research', *International Studies of Management and Organisations*, 12(4): 23–70.

Cummings, L. L. and Scmidt, S. M. (1972) 'Managerial Attitudes of Greeks: the Role of Culture and Industrialisation', *Administrative Science Quarterly*, 17(2): 265–272.

Deery, S. and Kinnie, N. (2004) 'The Nature and Management of Call Centre Work', in S. Deery and N. Kinnie (eds) *Call Centres and Human Resource Management: A Cross-National Perspective*, Hampshire: Macmillan.

Edwards, P. K. (1986) *Conflict at Work*, Oxford: Basil Blackwell.

Edwards, P. K. (1987) *Managing the Factory*, Oxford: Basil Blackwell.

Edwards, R. (1979) *Contested Terrain: The Transformation of the Workplace in the Twentieth Century*, London: Heinmann.

EIRR (European Industrial Relations Review) (2000) 'Call Centres in Europe', *European Industrial Relations Review*, 321: 13–20.

Epilogi (2003) 'The Leading Players in Telecommunications', May, p. 23 (in Greek).

Ferner, A. (1997) 'Country of Origin Effects and HRM in Multinational Companies, *Human Resource Management Journal*, 7(1): 19–37.

Ferner, A. and Quantanilla, J. (1998) 'Multinationals, National Business Systems and HRM: the Enduring Influence of National Identity or a Process of Anglo-Saxonisation?', *The International Journal of Human Resource Management*, 9(4): 710–731.

Fox, A. (1974) *Beyond Contract: Work, Power and Trust Relations*, London: Faber and Faber.

Frenkel, S., Tam, M., Korczynski, M. and Shire, K. (1998) 'Beyond bureaucracy? Work Organisation in Call Centres', *The International Journal of Human Resource Management*, 9(6): 957–979.

Guest, D. E. and Hoque, K. (1996) 'National Ownership and HR Practices in UK Greenfields Sites', *Human Resource Management Journal*, 6(4): 50–74.

Heckscher, C. (1994) 'Defining the Post-bureaucratic Type', in C. Heckscher and A. Donnellon (eds) *The Post-Bureaucratic Organisation*, London: Sage.

Hofstede, G. (1980) *Culture's Consequences: International Differences in Work-Related Values*, Beverly Hills: Sage.

Katz, H. and Darbishire, O. (2000) *Converging Divergences – Worldwide Changes in Employment Systems*, Ithaca: Cornell.

Kritsantonis, N. D. (1998) 'Greece: the Maturing of the System', in A. Ferner and R. Hyman (eds) *Industrial Relations in the New Europe*, Oxford: Blackwell.

Lane, C. (1989) *Management and Labour in Europe*, Aldershot: Ashgate.

Leidner, R. (1996) 'Rethinking Questions of Control: Lessons from McDonald's', in C. L. Macdonald and C. Sirianni (eds) *Working in the Service Society*, Philadelphia: Temple University Press.

Marginson, P. and Sisson, K. (1994) 'The Structure of Transnational Capital in Europe: the Emerging Euro-company and its Implications for Industrial Relations', in R. Hyman and A. Ferner (eds) *New Frontiers in European Industrial Relations*, Oxford: Blackwell.

Papalexandris, N. (1999) 'Global Leadership and Organisational Behaviour Effectiveness – Greece: from Ancient Myths to Modern Realities', paper presented at the *Second Globe Anthology*, Athens, December.

Sawyer, M. (1989) *The Challenge of Radical Political Economy*, Hemel Hempstead: Harvester.

Taylor, P. and Bain, P. (1999) 'An Assembly Line in the Head: Work and Employee Relations in the Call Centre', *Industrial Relations Journal*, 30(2): 101–117.

Taylor, P., Hyman, J., Mulvey, G. and Bain, P. (2002) 'Work Organisation and the Experience of Work in Call Centres', *Work, Employment and Society*, 16(1): 133–150.

The Economist (2001) '*Multinational Monitor*', 13 June.

Watson, T. J. (2003) *Sociology, Work and Industry*, London: Routledge.

Whitley, R. D. (1992) *European Business Systems: Firms, Markets and their National Contexts*, London: Sage.

11 Escaping the electronic birdcage

Workplace strategies in Swedish call centres

Antoni Lindgren and Per Sederblad

Introduction

The aim of this chapter is to discuss the possibilities for worker autonomy in relation to call centre work. Limited autonomy can be found in the practices of the daily work within call centres and some illustrative examples will be provided here from the case studies presented. Due to changes in market conditions and global competition, managerial strategies – including more radical changes towards increased autonomy in using technology and team organisation – are evident in these studies. As such, despite limited autonomy, there is a restricted freedom inside the electronic birdcage. Thus, it appears that for employees to find more substantial freedom in the workplace, there must be the possibility of leaving 'the cage'.

Theoretically, this chapter has its background in labour process analysis and Scandinavian working life research. On a theoretical level, the concept 'flexible autonomy' is introduced. For more advanced forms of work organisation changes see Friedman (1977) and Piore and Sabel (1984). The focal points of this chapter also relate to ongoing discussions concerning call centres and human resource management (see Bain *et al.* 2002; Deery and Kinnie 2004). The contribution from this chapter to these debates is based on case studies of 'the new service work' in Swedish call centres.

Workplace control and 'new service work'

Emotional labour and sense-making have historically been identified as fundamental for working conditions, with factors such as work organisation, stress, control and personal development evident in service work frequently related to the managerial strategies utilised. The concept of 'factory regimes' (Buroway 1985), once commonly used when researching working life, indicates the same notion – that managerial strategies are influential in terms of the overall working conditions, including relations at the workplace.

The 'new kind of service work' differs from the 'old service work' as personal services have moved from face-to-face interaction, taking place through new information and communication technology where telephones and computers are integrated. Yet, the basic function of the new service work is still similar to the old service work where there is a 'direct meeting' between the seller and the customer, taking place on the telephone or on the Internet within call centres.

A call centre is perceived as a workplace where telephones and computers are integrated and controlled by an expert system, and where work performance is frequently measured and electronically monitored. Thus, work performance tends to be strictly controlled through a managerial strategy of 'direct control' combined with 'technical control' (Friedman 1977; Edwards 1979).

In this chapter the authors outline how one of the studied call centre workplaces evolved from direct control towards flexible autonomy for employees. To describe this shift in working conditions to 'flexible autonomy' it is useful to make a distinction between the work performance, that is, the labour process, and the workplace, or that part of the organisation which forms the physical surroundings of the labour process.

In an analogous way to Friedman (1977), who introduced 'responsible autonomy' as an alternative to 'direct control', and Piore and Sabel (1984), who identified 'flexible specialisation' as an alternative to mass production, the authors propose that there is a 'second divide', also evident in service work, which is 'flexible autonomy'. The meaning of flexibility here is that the organisation should be capable of adjusting as required to the market and customer companies. Autonomy can be reached on an individual level and also on a team level, where a team has to work with a specific customer, company or a set of companies.

'Technical control' and 'direct control' are elements of the typical working conditions at call centres, and have been reported in many studies, especially from the UK (Taylor and Bain 1999; Callaghan and Thompson 2001). An emphasis on control often provides problems with the quality of the services. The suggested employer solution to this problem is often 'more control', that is more 'technical control' employed through the direction of employees, 'scripting' their work performance using the expert systems designed for this purpose and through more 'direct control', listening in on their calls and using test callers and similar (Bain *et al.* 2002).

Conversely, in the call centres that we studied, the management strategies which provided autonomy at the workplace also provided some degree of freedom in work performance. In one case, a call centre had an answering service for companies as their business idea. 'Your face to the customer' was their slogan, promising both 'effectiveness and friendliness'. In the case of this call centre, they had relatively low call quotas, with some 300 calls per day being the norm, compared with other call centre quotas (Taylor and Bain 2001). This centre had also introduced 'team organisation', mainly as a method to strengthen normative identification with the company (Thompson and Wallace 1996; Sederblad 2004). Moreover, the call centre managers provided considerable freedom in this workplace, with fringe benefits such as free tea and coffee, a pleasant dining room and the ability to take breaks when they wanted to. It appears that these managers understood some of the inherent conflicts between 'effectiveness and friendliness' and acted accordingly.

Before turning to the presentation of the method and results from the call centre case studies, the relationship between resistance and workplace strategies should be discussed. Specifically, flexible autonomy rests on a workplace strategy that gives freedom in the workplace, compensating for the strictly technically controlled labour process of the work performance. It has been noted in the UK call centre studies that a strictly controlled work performance, that is, scripting and the provision of limited employee freedom, can lead to them becoming agitated and creating resistance (Bain and Taylor 2000; Houlihan 2004).

A workplace strategy refers to a strategy which management uses as a response to, or through, an awareness of workplace resistance. The workplace strategy may be part of a management strategy, based on direct and technical control, which predominates in mass production call centres. Thus, 'resistance' and the quality of the employee–customer relationship work in favour of alternative forms of management, for example, the use of flexible autonomy and in permitting employees to escape the electronic birdcage.

Methods

The research approach taken here is hermeneutic. By combining empirical results gathered from interviews, observation and surveys with concepts and theories referring to working life research, the authors have gained an understanding of the labour process in new service work. The call centre research was supplemented by case studies of work in the travel agency industry, an industry which is also going through major structural changes in the way work is performed (Lindgren and Sederblad 2005).

There are three cases, which are referred to as CC1 (the smallest), CC2 and CC3 (the largest). CC1 was followed for five years. The long duration of this case study made time, observation and reflection, in relation to the changes taking place, important for our understanding of the changes in the labour process. Most of the interviews and the observations are drawn from CC1. The survey was conducted at CC1 and CC3. At CC2 we made some observations and conducted several interviews. All three call centres were owned by the same organisation, CC1 and CC2 were linked, while CC3 was a separate subsidiary.

As with any other organisation, a call centre is embedded in many respects. For example, in the society (local or national or global), in politics (neo-liberal privatisation or socialist regulation) (Wolf 2005) and in relation to the owners (whether they are national or international). A call centre also has its own historical traditions and managerial trends. In the case of call centres, technological development has radically changed the 'rational' options (Dosi 1988; Lindgren 1990) available in service work. In sum, the working life, and working life research, forms a veritable museum of ideas and practices in terms of organisation and management. Thus, through a hermeneutic approach one can actually 'read' the managerial strategies that have been adopted via the artefacts of the workplace. In the workplace one can also read into the social embeddedness of the organisation. In the case studies referred to in this chapter, one can also read the generic character of the new service work, that is, the emphasis on the 'customer relationship' as still the basic relationship. In spite of the new electronic technology functioning as a birdcage with regard to the performance of work, technology can, at the same time, provide possibilities for freedom. As there is always a choice between the application of different managerial strategies in organising the labour process and the workplace, there are possibilities for autonomy to occur in the context of new service work. Understanding the possibilities of such autonomous working conditions presupposes an understanding of the motives of management, such as being conscious or otherwise of the 'rational' decisions behind such managerial strategies.

Results

The average size of call centres in Sweden is around 300 employees (HTF 2000) which is a low figure compared with call centres in other countries (Taylor and Bain 2004). Although there are deviations from the average, these case studies are located at the small end of the scale. The three call centres in this study were all 'outsourced', independent companies and were all located in the far northern part of Sweden. CC1 had some 30 employees and CC1 was integrated with CC2, having approximately 90 employees, situated in another town. Both call centres had a telephone answering service as their main service. CC3, which was located in a different town, had 130 employees. They provided ticket services for a large airline as their main service and also had some telemarketing services. However, CC3 was closed down in 2003 due to a conflict between management and employees.

CC1 was integrated with CC2 and differed only in relation to its small size and business hours, which were those of a 'normal business'. We returned to this call centre a couple of times every year during the five-year period between 1999 and 2003 and, in that way, found that 'time' became an important part of our investigation. Many call centres do not experience longevity. Hence, by following one call centre for several years the authors were provided with an opportunity to discover what is important for determining the quality of working conditions.

In the office landscape the telephone operators were situated as three and three, facing each other, in a way that seems to be common for call centres. Each of the operators had a personal computer and a telephone headset. The telephone was integrated into the expert system and when there was an incoming call, that company's name would appear on the screen, along with the phonetic pronunciation. The tone used to give answers – cordial, neutral or cheerful – as well as the level of the telephone service and what information (such as mobile telephone numbers) that operators were allowed to give was also presented on the computer screen for operators to see. In both parts of the room, situated so that everyone could see it, was a monitor indicating the status of the service given both in terms of the length of the queue of incoming calls and the average answering time. Calls were to be answered within the number of signals being contracted and when there was a risk of this not being achieved the two staples turned red instead of green. This information was also displayed in numbers on each computer screen.

The quota that the call centre operators had to uphold was 300 calls per day. In addition to this, they had to take text messages, which resulted in extra wages for them. This was easily achieved by asking if the caller wished to leave a message for the person being called who may not have been available. The telephone operators said that there was no problem completing this quota, and the representative of the union also thought it was achievable.

The overall impression of the workplace was that it was light, that the air conditioning worked well and that the sound level was low, thanks to the sound-absorbing boards in the roof. The worktables and chairs were ergonomically fitted to operators. Employees also appeared to be content with their physical work environment. But although CC1 was a good place to work, there was also discontent. Why? During 2000 there was discontent with the working time schedule. There was a sharp shift in the atmosphere at the call centre, from positive to negative. Management had introduced a new computerised time schedule that was operational for only one week at a time. This time frame did not enable any long-term scheduling of work.

The discontent went on for one year, the time it took to solve the scheduling problem and to introduce a more long-term work schedule. Management then introduced work groups, causing a new round of discontent. Employees were grouped together according to principles of which they were not aware. They were also told to have group meetings once a month and to rotate the group leadership. The union thought that it was a way to avoid informal groupings, and the team leader, as the representative of management, said it had something to do with raising the level of calls handled by the weakest of the employees. As researchers, we never did receive a satisfactory explanation and understanding of the introduction of these work groups. In 2003 the work groups only remained as a kind of leisure group. A small amount of money was connected to the leisure activities of the group, but the employees still had to sit in these groups regardless.

'Friendliness and effectiveness' was one of the arguments used by the company while selling the answering service to other companies. 'We will be your face towards the customer', was another. Both these arguments rest on the division between emotional labour and emotion work (Hochschild 1983). When individuals have to work with their feelings in a forced way – be nice, cheerful and so on, even though they do not feel that way at all – they are forced to provide emotional labour. The answering service rested mainly on emotional labour, but it seems that if there is to be any quality in the service, there must also be some possibility for emotion work. We were able to listen in on the calls taken, sitting side by side with the telephone operators and observe how they handled this demand for emotional labour. First of all we felt the tempo to be high and the work task demanding. Concentration is required for this kind of work, as well as speed and diplomacy. Most of the customer companies oper-ators were serving were located in the Stockholm City area – stock exchange brokers, real estate brokers and telecom operators. The oper-ators never knew who would be on the other side of the line, whether it would be a pleasant person, a rude one or someone in distress, so they had to be emotionally prepared to handle any kind of call. We noticed that the older telephone operators, those who had been employed for at least a couple of years, had some techniques for handling all of these varied situ-ations. The 'star operators' were also very efficient, handling a call in almost no time at all and at the same time being polite. The young ones were more cheerful, not yet having learned how to save their emotional energy. At one time when we were sitting listening in with one of the senior telephone operators she said: 'sometimes you feel like a parrot!' We had been listening in for more than an hour and not noticed any stereo-typed answers, but still she felt this way – like being in a birdcage!

CC2 had some 90 employees and the overall impression was that every-thing was 'bigger' than it was in CC1. The building was bigger, not easy to overlook since it was divided into many rooms and corridors. There was a main hall, where most of the telephone operators were located, and adjoining this hall were several small rooms hosting customer relations, technical support and the like. Management were located in Stockholm, far away, but some managers spent a day or two a week at this plant and had their local offices adjoining the main hall. Thus, there was a concentra-tion of operators to one work area and managers were located in other areas. At the work area, the operators were sitting three and three in the same way as in CC1. There was a difference though, since there were a few customer groups in the larger call centre, just as they had been introduced in the business travel agencies.

CC3 had some 130 employees and was located in a two-storey building in an industrial area where there were also other large call centres. Two-thirds of the employees worked with ticket sales for a large airline, using their software. In the autumn of 2001 we conducted interviews at CC3 and also conducted a survey of employees' work tasks. Some employees had been working for two airlines before and were practically doing the same work, but within a call centre. One-third of the employees were in tele-marketing, located on the ground floor; ticket sales and management were located on the first floor. They were all very young, around 20 years old, and sat three and three in the same way as at the other two call centres. Those upstairs, though, sat one and one and had screens marking their own work area, indicating a more 'private' work environment. In fact they sat in the same way as they do in business travel agencies, where the employees have their own work desk. They were also more formerly attired. Downstairs, on the other hand, employees were very relaxed, sitting laid back in their chairs, chatting in between calls, drinking soft drinks and dressed very informally and often with body decorations. Upstairs it was crammed and dark, downstairs it was roomy and light. The overall impression, though, was that the physical work environment was as good as it was in the two other call centres.

In 2002 management tried to introduce split shifts and new work time arrangements; the employees would work a few hours in the morning and then come back in the afternoon or evening and work some more hours. When the employees refused to do this they were fired and the plant was closed down by management. Evidently, even a relaxed work environment can become tense and management directives can be resisted by the most relaxed of workers.

Discussion

CC1 was more intensively researched over a longer period of time than the other two call centres. From this study we gained an understanding of what is fundamental for this kind of new service work, besides the ephemeral changes which also took place. Among the latter kind of changes, were the work groups; we never really found any explanation why they were introduced and neither, it seems, did the local management, the employees or the representative of the union.

The starting point of our understanding was that autonomy is essential to everyone and, therefore, very strictly controlled labour processes lead to agitated people. This, in turn, calls for a different type of workplace strategy in order to secure quality of the work performance. This is necessary due to the demands evident from employee–customer relationships within call centres where managers wish to secure the emotional labour of the employees. When we commenced our study there was discontent with the work time schedule. After that, when this problem had been solved, there was discontent with the work groups, which never completely disappeared. There was also an atmosphere of ongoing, but vaguely defined, dissatisfaction evident amongst employees. We believe that this discontent and vague dissatisfaction had to do with a need for autonomy. Research into working life, ever since the Mayo studies in the 1930s, has shown that respect, which presupposes autonomy, makes people work better and can act as a motivator.

The fundamental conditions of work at CC1 were expressed not only in a number of the artefacts representing the 'workplace strategy' but also through resistance and the quest for freedom by the call centre operators. The most important cultural artefacts were: the work time schedule, the coffee percolators, the fax machine room and the dining room.

During our first visit to CC1 we noticed that the first thing the telephone operators did when they came to their workplace in the morning was to check the day's work time schedule. They checked their coffee breaks, lunch breaks and the time they would finish work. Thus, they did not know the schedules in advance, although they could find out. Why then did they become highly discontent when management introduced the new one-week work roster scheme? We 'read' it in this way. The operators' discontent had to do with freedom – or lack of it, that is, it is important that you know when you have to be at work, and when you are free, in order to make other arrangements and organise your life away from the workplace. Of course, there must have been differences among the telephone operators in the way they used the work time schedule. The operators observed for this study were mostly young and unlikely to have dependants. The atmosphere became friendlier after yet more new schedules were introduced. As a result, there was even more freedom than before, since employees started a new schedule by choosing their own working hours and afterwards the team leader made adjustments according to the needs of the plant, discussing the adjustments with employees on an individual basis.

The coffee percolators were probably the most important artefacts because they were so conspicuous. In Sweden people normally drink coffee at their breaks – thus it is important how it tastes. But there was also a coffee machine, of the kind which catering firms provide where you can choose between many different kinds of coffee and chocolate. Still employees used the percolators: Why? Were they not redundant? No! We 'read' these actions in this way. The coffee machine was management's device. Instead, the employees brewed their own coffee, thereby manifesting their freedom and their resistance.

What about the dining room? It was run by the telephone operators themselves. They had a work schedule devised by them outlining who should take care of the dishwasher, manage the refrigerator, undertake the cleaning and arrange that tea, coffee and biscuits were purchased. It was their dining room and it gave employees autonomy and freedom.

What about the fax room? We 'read' it as representing something private. This was the only place where they could make a telephone call that was not registered – they could also make calls from the dining room if they wished but then it would not be totally private – or send emails that were not registered, nor electronically scanned. Management was very cautious about listening in on calls; they used this only during the training period at the beginning of an employee's contract. However, the team leader spent every morning gathering individual and plant statistics for the previous day, so there was still an awareness of the electronic surveillance.

Conclusions

Freedom was evident at the CC1 workplace and this was part of management's strategy. The work performance and the labour process were strictly controlled on the one hand, but compensated for by freedom at the workplace and by a workplace strategy. A lower quota of calls than was normally found in call centres, the creation of work groups and the discussions about team formation were part of the flexible autonomy provided to employees. The same situation applied to the possibility for employees to take a few minutes break at their own will. Nevertheless, the labour process was strictly controlled. The workplace strategy thus worked as a complement to the organisation of the labour process. It gave compensation for the lack of freedom in the work performance, making it easier to put up with it. It also temporarily provided an escape from the electronic cage. Flexible autonomy represents an alternative way of managing this kind of service work to the 'factory regimes' that still dominate in call centres. In these cases, we have identified how management can use flexible autonomy strategies within Swedish call centres. As can be seen in the case of the large call centre, sometimes such strategies may be necessary if they want to stay in business.

References

Bain, P. and Taylor, P. (2000) 'Entrapped by the 'Electronic Panopticon'? Worker Resistance in Call Centres', *New Technology, Work and Employment*, 15(1): 2–18.

Bain, P., Watson, A., Mulvey, G., Taylor, P. and Gall, G. (2002) 'Taylorism, Targets and the Pursuit of Quantity and Quality by Call Centre Management', *New Technology, Work and Employment*, 17(3): 170–185.

Buroway, M. (1985) *The Politics of Production*, London: Verso.

Callaghan, G. and Thompson, P. (2001) 'Edwards Revisited: Technical Control and Call Centres', *Economic and Industrial Democracy*, 22(1): 13–37.

Deery, S. and Kinnie, N. (eds) (2004) *Call Centres and Human Resource Management – a Cross-national Perspective*, Basingstoke: Palgrave.

Dosi, G. (1988) 'The Nature of the Innovative Process', in G. Dosi (ed.) *Technical Change and Economic Theory*, UK: Panther Publishers.

Edwards, R. (1979) *Contested Terrain. The Transformation of the Work Place in the Twentieth Century*, New York: Basic Books.

Friedman, A. (1977) *Industry and Labor: Class Struggle at Work and Monopoly Capitalism*, London: Macmillan.

Hochschild, A. R. (1983) *The Managed Heart. Commercialisation of Human Feelings*, USA: University of California Press.

Houlihan, M. (2004) 'Tensions and Variations in Call Centre Management Strategies', in S. Deery and N. Kinnie (eds) *Call Centres and Human Resource Management – a Cross-national Perspective*, Basingstoke: Palgrave.

HTF (Salaried Employees' Union) (2000) 'Den "nya" arbetsmarknaden' (The 'New' Labour Market) Report, Stockholm.

Lindgren, A. (1990) 'Strategimodellen – en modell för förståelse av industriell utveckling' (The Strategy Model – a Model for Understanding of Industrial Development), in Håkon Finne (ed.) *Fra redskap till budskap*, Trondheim: IFIM.

Lindgren, A. and Sederblad, P. (2005) 'Arbetsförhållanden i call centers och resebyråer' (Working Conditions in Call Centres and Travel Agencies), Research report, Stockholm: AFA.

Piore, M. J. and Sabel, C. F. (1984) *The Second Industrial Divide. Possibility for Prosperity*, New York: Basic Books.

Sederblad, P. (2004) 'Editorial Introduction to Special Issue on New Forms of Teamworking', *Economic and Industrial Democracy*, 25(2): 187–196.

Taylor, P. and Bain, P. (1999) '"An Assembly Line in the Head": Work and Employee Relations in the Call Centre', *Industrial Relations Journal*, 30(2): 101–117.

Taylor, P. and Bain, P. (2001) 'Two Steps Forward, One Step Back: Interest Definition, Organisation and Dissipated Mobilisation Amongst Call Centre Workers', paper presented at the *19th Annual International Labour Process Conference*, Royal Holloway, University of London, March.

Taylor, P. and Bain, P. (2004) 'Call Centres in Scotland and Outsourced Competion from India', Research Report, University of Stirling/University of Strathclyde.

Thompson, P. and Wallace, T. (1996) 'Redesigning Production through Teamworking', *International Journal of Operations and Production Management*, 16(2): 103–118.

Wolf, M. (2005) 'More Public Spending does not Lead to Slower Growth', *Financial Times*, 23 March: 15.

Index